D0982619

W

THE INTERNET IS N

WHAT YOU THINK I

The Internet Is Not What You Think It Is

A HISTORY, A PHILOSOPHY, A WARNING

JUSTIN E. H. SMITH

PRINCETON UNIVERSITY PRESS
PRINCETON & OXFORD

Published by Princeton University Press
41 William Street, Princeton, New Jersey 08540
6 Oxford Street, Woodstock, Oxfordshire OX20 1TR

press.princeton.edu

Library of Congress Control Number 2021950463
ISBN 978-0-691-21232-6
ISBN (e-book) 978-0-691-22968-3

British Library Cataloging-in-Publication Data is available

Editorial: Rob Tempio and Matt Rohal
Production Editorial: Jill Harris
Jacket Design: Karl Spurzem
Production: Erin Suydam
Publicity: Maria Whelan and Carmen Jimenez
Copyeditor: Karen Verde

Jacket image: Mycelium, *Fusarium euwallaceae.* Courtesy of Protasov AN / Shutterstock

This book has been composed in Arno

10 9 8 7 6 5 4 3 2 1

Now this connexion or adaptation of all created things to each and of each to all, means that each simple substance has relations which express all the others, and, consequently, that it is a perpetual living mirror of the universe.

—G. W. LEIBNIZ (1716)

[T]hose who . . . think on the mathematical truth as the instrument through which the weak mind of man can most effectually read his Creator's works, will regard with especial interest all that can tend to facilitate the translation of its principles into explicit practical forms.

—ADA LOVELACE (1843)

This is the struggle with describing social media: it devours importance.

—LAUREN OYLER, *FAKE ACCOUNTS* (2021)

CONTENTS

ACKNOWLEDGMENTS

MANY DIFFERENT PEOPLE provided many different kinds of support and insight leading to the completion of this book. I suppose I must first thank the editors at *The Point Magazine*, for publishing my essay, "It's All Over," in January 2019. This essay, the closest thing to a viral hit I've ever produced, would be the germ from which the present book grew. For their helpful comments on various iterations of this book's drafts, I heartily thank Agnes Callard, James Delbourgo, Yves Citton, Michael Friedman, Jonardon Ganeri, Ohad Nachtomy, Jessica Riskin, Becca Rothfeld, Eric Schwitzgebel, and Galen Strawson. I am lucky to work with so many wonderful colleagues and students in the Laboratoire Sphère of the Université de Paris, whose tremendous gamut of research interests provides me with a steady continuing education and a constant stream of new insights and considerations for my own work; thanks especially to Claude-Olivier Doron, Jean-Baptiste Grodwohl, Vincenzo De Risi, Karine Chemla, and Jean-Jacques Szczeciniarz. I am immensely grateful to the Cullman Center for Scholars and Writers, where I held the John and Constance Birkelund Fellowship in 2019–20 and where I made tremendous progress on this book, even though, deviously, it was not the book I was invited there to complete (that one is on its way, I promise). I thank, in particular, the director of the Cullman Center, Salvatore Scibona, as well as my fellow fellows, Ken Chen, Hua Hsu, and Sally Rooney, and Eric Sanderson of the Bronx Zoo, for his contributions to my understanding of ecology and of the limits of ecological metaphors for thinking about the internet. I am no less grateful to my collaborators in the 2021 workshop in the history of science at Princeton University on the theme of "Attention: History, Science, Philosophy," notably to my co-organizer D. Graham Burnett, and also to Carolyn Dicey Jennings, Jesse Prinz, John Tresch, Natasha Dow Schüll, and all the other participants. I extend special thanks to all the brave pioneers who participated in the open and experimental Zoom seminar that I conducted on Kant's *Critique of the Power of Judgment* from March to May 2020, while under quarantine in New York and suffering from COVID-19 and its after-effects (another sort of viral hit). I thank in particular Ognian Kassabov, Jake McNulty, Catherine Wilson, and Johanna Winant for their contributions. That seminar was an absolute lifeline for me, and its impact on my thinking is evident on nearly every page—not only the ones that discuss Kant. Special thanks also to Augustus Amund Wellner, for sharing my taste in memes. And thanks finally, as always, immeasurably, to Adina Ruiu.

THE INTERNET IS NOT WHAT YOU THINK IT IS

Introduction

"Let us calculate!"

"TO STRENGTHEN our social fabric and bring the world closer together." This, maintains Mark Zuckerberg, CEO of Facebook Corporation, is his enterprise's reason for being. Yet it would not take a particularly critical mind to notice that strengthening the social fabric and bringing the world closer together are not, in fact, what Facebook is doing. No, Facebook and the other big tech companies are, plainly, tearing the social fabric to threads, and pulling people apart.

Just fire up your computer and marvel at the news of the day, at all the angry people behind the avatars, fighting with one another and with bots about the news, and about the meaning of the news. Witness how global and local politics have been corrupted into a form of unrelenting disinformation warfare. See organized trolling campaigns fomenting violence against minority groups throughout the world. Observe the mobbing of political dissidents by mass campaigns from below, and the repression of the same dissidents by state surveillance from above. Revisit 2016 and watch the technologies of the new big tech companies mobilize to propel a disreputable internet troll into the highest office of the most powerful country in the world. Tremble before the online rage addicts who daily band together in search of new targets: someone caught on video in a moment of indiscretion, who is then summarily doxxed (that is, has their personal information revealed on the internet), shamed, fired, or ostracized; some young adult on the cusp of success who is brought low when shown to have used hateful language as a teenager in a chat forum; some clueless normie (slang for a normal, mainstream person, oblivious to the rhythms and insiderisms of online culture) ruthlessly ridiculed for not yet having adopted the terminology for a given identity group that was ratified by social-media vanguards only a short time before. There is no sign that anyone has a clear plan, or the necessary power, to abate the chaos these technologies have unleashed. We are living in a crisis moment of history, in the true sense of "crisis": things might get better eventually, but they will never be the same.

As recently as ten or fifteen years ago, one could still sincerely hope that the internet might help "to bring people together and to strengthen the social fabric." When the revolutions of the so-called Arab Spring began to break out, many of us, myself included, declared that this was the power of social media being unleashed, hailing a new era of democracy and egalitarianism throughout the world.

The arc of such utopian hopes is long, and it has decisively bent in the direction of defeat only in the last decade. The dream of a rationally governed society, freed of passionate human conflicts through the outsourcing of decision-making procedures to machines, is one that the German philosopher Gottfried Wilhelm Leibniz already articulated as early as the 1670s. In a text in which he develops an artificial and formal language for the exact expression of all natural-language terms, the philosopher envisions a near future in which, "if controversies were to arise, there would be no more need of disputation between two philosophers than between two calculators. For it would suffice for them to take their pencils in their hands and to sit down at the abacus, and say to each other: *Let us calculate!*"[1] The "abacus" in question is not a real abacus, but any tool that might aid in processing the formal language, though in principle Leibniz also thinks, as he conveys in this passage, that the language can be deployed using only a pen and paper (just as one might do long division either by hand or by using some sort of calculator).

This hortatory third-person-plural use of the Latin verb "to calculate"—*Calculemus!*—might well serve as the motto of Leibnizian optimism, of the belief that all problems can be resolved simply by clarifying our terms and rationally following the logical consequences of our commitments. This optimism extends not just to disputes between philosophers arguing over abstractions about the nature of substance or the immortality of the soul, but also to diplomats representing empires on the brink of war. For Leibniz, the development of a universal formal language is a key part of the imminent attainment of world peace, a part that would continue to capture imaginations in a more demotic form well into the twentieth century, where artificial languages such as Esperanto, Volapük, and Ido often appealed to peace activists of various strains, some of whom, notably Bertrand Russell (an advocate of Ido), also owed a deep philosophical debt to Leibniz.[2]

The history of artificial languages and the history of computing go hand in hand, and while the reckoning engine that Leibniz developed (which we will discuss on several occasions below) was only intended for arithmetical calculations, he well understood that in principle such a machine could also be used to process any information at all. In part this understanding was deepened by his important contributions to the development of the binary calculus, which makes it possible to encode any proposition in a sequence of zeroes and ones, and thus to process language using the same tools with which one might also process numbers. In part, Leibniz's awareness of the possibility of concept-crunching machines, and not just of number-crunching machines, came from the fact that he was working in an already centuries-long tradition of thinking about such devices, some of which were merely fantastical, some of which may have actually existed.

Thus, in the early fourteenth century, the Majorcan polymath Ramon Llull designed a machine made of paper, consisting of several concentric discs marked with symbols on the edges denoting various attributes of substances. By rotating these discs one could, Llull hoped, exhaustively survey all of the combinatoric possibilities for the kinds of being in (and beyond) the world. Leibniz took Llull as an important predecessor in the history of formal-language processing, and Llull had his own influential predecessors too, notably Aristotle, as well as other sources in the Jewish and Islamic mystical traditions of Al-Andalus. While we might be tempted to see Leibniz, perhaps along with his contemporary Blaise Pascal, as the "father" of computer science, in truth computers have no father, or mother, and for any starting point you might attempt to choose in history, you can always find other predecessors with whom the thinker standing at that starting point was already in conversation, to whom that person was responding, who served as their starting points.

What happens with Leibniz is not the proper beginning of anything, but rather—a metaphor to which we will be returning frequently—it is the *weaving together* of several ideas into a filament thick enough to serve further on as a bright guiding thread through the rest of modern history up to the present course not yet called by that name). In the early 1960s, Norbert Wiener was sharply aware that the possible apocalyptic results of modern technology might result simply from our loss of control over machines to which we have outsourced decision-making processes, and thus to teach a machine to play chess may already give it more responsibility than it can handle over war, peace, and human destiny. "There is nothing more dangerous to contemplate than World War III," Wiener writes in a supplementary chapter of the second edition of his *Cybernetics*, to which we will be returning throughout this book.[3] And, he adds: "It is worth considering whether part of the danger may not be intrinsic in the unguarded use of learning machines."[4]

A general wariness of modern technology pervades much mid-twentieth-century existential and phenomenological philosophy, frequently, as in Martin Heidegger, with discomforting undertones, and sometimes outright explicit claims, of the conflict between technological enhancement of our social lives, on the one hand, and "authentic" living on the other. This pessimism continues to echo in late twentieth-century psychological, psychoanalytic, and social-scientific engagement with the problem of modern "alienation" and the ways in which technological enhancements remove us from the human bonds and natural attachments that make life meaningful. In the 1970s, sociologists such as Manfred Stanley warned against the rise of "technicism" in interpreting human actions and motivations, and in so doing were criticized by others for their "pessimism." Yet like Stanley, and unlike Heidegger or some caricature of the "Luddite," I am interested here in "eschewing apocalyptic frenzies of doom or salvation in favor of calmer analysis."[5] While strongly opposed to the "technicist mystification of personal consciousness under conditions of modern industrial civilization"[6] and concerned to salvage "human dignity" under these conditions,[7] I am likewise concerned to show that the greatest problem is not one of unstoppable technological determinism, or of a determinism that can only be countered by "flipping the off switch," but rather in clarifying the nature of the force with which we are contending, and understanding the limits of thinking that proceeds by analogy between human beings and machines. Stanley's approach is largely through the analysis of language, while mine is through history, but in both cases the aim is to engage in lucid criticism while avoiding the pitfalls of pessimism or authenticity-mongering.

———

I have been using the term "internet" in an overtly non-technical way. The internet, after all, is the entire network of networks that are connected by the Internet protocol suite. The "World Wide Web" that we commonly access through our familiar browsers is only one small part of this network. And the sites that will be of principal interest for us in the pages to follow are only one small part of what may be accessed on the World Wide Web. I am not centrally concerned, here, with the social implications of our new ability to access, say, digitized medieval manuscripts held by the Bibliothèque Nationale in Paris (though such new possibilities do become the center of attention in chapter 5), but with the more familiar sites of daily use by billions of people: Facebook, Google, and so on. Thus, "internet" serves as a sort of reverse synecdoche, the larger containing term standing for the smaller contained term. The reason for adopting this terminology is that it seems to agree with actual usage among current English speakers; on Twitter, for example, you will often see users declaring exasperatedly that their antagonists need to "get off the internet" and "touch grass." Here, they don't really mean the whole internet; they mean Twitter.

To put this another way, I am concerned with the "phenomenological internet": the one we know directly through its appearances to us, and the one we commonly describe by that name; I date my own first use of "the internet" to a certain day in 1997, which was the first time I saw an html-based homepage, though I had been sending e-mails for five years before that and had connected to the networked computer service known as "The Source," on my father's old Kaypro, plugging our landline phone into an auxiliary suction-cup modem, as early as 1980. It seems reasonable terminologically to follow actual usage, and it seems conceptually justified to focus on the small corner of the internet that is phenomenologically most salient to human life, just as when we speak of "life on earth" we often have humans and animals foremost in mind, even though all the plant life on earth weighs over two hundred times more than all the animals combined, in terms of total biomass. Animals are a tiny sliver of life on earth, yet they are preeminently what we mean when we talk about life on earth; social media are a tiny sliver of the internet, yet they are what we mean when we speak of the internet, as they are where the life is on the internet.

Let us imagine, if we are able, a not-so-distant future in which the internet, or some suitable representative of this diffuse entity, finds itself in the dock, under prosecution for all the harms it has unleashed upon our fragile world. Let us not focus on its minor transgressions, the particular industries it has killed off or is threatening to kill off: journalism, music, film, higher education, publishing. In such cases we are only seeing what the tech enthusiasts like to call "disruption," of the sort we see after the introduction of any new technology. And just as photography disrupted various practices, including book illustration, portraiture, and so on, without ultimately stunting or limiting our ability to represent the world around us, so too, for nearly every human practice threatened by the internet, we are already witnessing exciting and promising new practices that expand rather than shrink our potential. Newspapers for example were good in their day, but there is nothing about the electronic dissemination of news that is in principle incompatible with the social good once provided by a trusty old broadsheet.

The principal charges against the internet, deserving of our attention here, instead have to do with the ways in which it has limited our potential and our capacity for thriving, the ways in which it has distorted our nature and fettered us. Let us enumerate them.

First, the internet is addictive and is thus incompatible with our freedom, conceived as the power to cultivate meaningful lives and future-oriented projects in which our long-term, higher-order desires guide our actions, rather than our short-term, first-order desires. Second, the internet runs on algorithms, and shapes human lives algorithmically, and human lives under the pressure of algorithms are not enhanced, but rather warped and impoverished. To the extent that we are made to conform to them, we experience a curtailment of our freedom. Third, there is little or no democratic oversight regarding how social media work, even though their function in society has developed into something far more like a public utility, such as running water, than like a typical private service, such as dry cleaning. Private companies have thus moved in to take care of basic functions necessary for civil society, but without assuming any real responsibility to society. This, too, is a diminution of the political freedom of citizens of democracy, understood as the power to contribute to decisions concerning our social life and collective well-being. What Michael Walzer said of socialism might be said of democracy too: that "what touches all should be decided by all."[8] And on this reckoning, the internet is aggressively undemocratic. Fourth, the internet is now a universal surveillance device, and for this reason as well it is incompatible with the preservation of our political freedom.

I shall have more to say about some of these indictments than others; in particular I am most interested in the first of them, the addictive power of the internet, which is one dimension of what we may call "the crisis of attention." But they all overlap in complex ways: increasingly, for example, social-media behavior in the form of likes for certain songs or artists, which might only have come to one's attention as a result of algorithmic processes over which one has no say, can also in turn place a person on the radar of law enforcement agencies or state security apparatuses as a potential terrorist, gang member, or other species of socially disadvantaged undesirable.

All of the major charges are related to one another, moreover, in contrast to the minor charges we are passing over concerning the destruction of this or that industry or art form, in that they involve, again, a threat to human freedom. Freedom is a difficult concept, in part because there are many different species of it. A Uighur in a Chinese detention camp, or a migrant in Texas with an ICE ankle monitor, is unfree, and so, in a different but somewhat related sense, is a hiker whose leg is caught under a fallen tree. A heroin addict is unfree in yet another distinct but related sense, and so are a wage laborer, a lay-about so entranced by soap operas as to never realize innate human potentials, and anyone else at all who, because of either inner weakness of the will or objective outer forces, fails in some way to become what they could have been, fails to achieve full human thriving. We are all unfree in some of these respects. The charge here is that the internet contributes to the limitation of freedom in all of these respects. As such, the internet is anti-human. If we could put it on trial, its crime would be a crime against humanity.

Things were not always expected to turn out this way. Figuring out what went wrong will be the principal concern of this book. But in order to do this, we will need to think deeply not just about the past few years of what the internet has wrought in politics, culture, and economics. This ground has been well covered by many lucid scholars and critics. We will rather need to focus on what the internet is, ontologically speaking, on the nature of this new thing we already so easily take for granted; and we will need to focus on what the internet is genealogically speaking, too, on its place in the vast sweep of human and even natural history. Only in so doing can we begin to see what the internet might yet become.

A few words are in order concerning "methodology." This book will strike some readers as peculiar, in that it purports to be a "philosophy of the internet," yet spends most of its time dwelling on thinkers, texts, and problems from centuries ago. This is intentional; this *is* the methodology. I am, by training, a historian of philosophy and science, with a particular long-standing interest in the intersection of philosophy and the life sciences in Europe in the seventeenth and eighteenth centuries, and with an abiding interest as well in philosophical aesthetics and the many points of contact between philosophy, science, and art throughout history. I also have a strong sympathy for some dimensions of the work of Michel Foucault, who well understood that some problems are best studied genealogically, that is, that we come to understand the essence of a thing by understanding how it develops over the course of history. This is thus in some respects a contribution to the genre of scholarship that Ian Hacking has called "historical ontology,"[9] that, namely, regards history as of central importance in any effort to understand what there is in general, or what the nature of a given thing that is, is. Thus, for example, if you want to offer up a "philosophy of cinema" (such an antiquated undertaking!), if you want to give an account of what cinema in its essence is, you must spend a good deal of time considering such things as nineteenth-century shadow plays and the narrative techniques of novelists such as Balzac or Flaubert.

Unlike Foucault, however, I am less inclined here to assent to the idea that different historical epochs are characterized by their own, radically distinct "epistēmēs." Indeed, my argument about the history of technology points much sooner in the opposite, perennialist direction: notwithstanding the enormous changes in the size, speed, and organization of the devices we use from one decade or century to the next, what these devices are, and how they shape our world, has been substantially the same throughout the course of human history (and, as we will see, even longer than that). So the book amounts to a kind of reverse Foucauldianism, or, if you will, a perennialist genealogy: bringing history to bear on a thing important enough to warrant philosophical attention, and determining through this historical-philosophical inquiry that the thing is more or less stable across the ages, and not a discursive product forever trapped within the confines of a single epoch's epistēmē, even if the current epoch does present us with some truly novel challenges.

In this short book we will range widely in topic and time, permitting ourselves to linger far from some of the questions that internet users and tech analysts today consider most pressing: the outsized power of the tech monopolies; the racism built into AI applications in security, social media, and credit-rating algorithms; the variations on the trolley problem to which self-driving vehicles give rise; the epidemic of disinformation and the corollary crisis of epistemic authority in our culture; internet mobs and the culture wars; and so on, ad nauseam. For the most part, this aloofness is intentional. This book does describe itself as a "philosophy" of the internet and, while there will be much disagreement about what that might mean, most of us can at least agree that a philosophy of something, whatever else it may be, has the right to zoom out from that thing and to consider it in relation to its precedents, or in relation to other things alongside which it exists in a totality.

But let us not suppose that zooming out can hold no practical lessons for the present day. Such an assumption is in part how we got into this whole mess in the first place. By treating the internet as a short-term problem-solver, we created for ourselves some new, very big problems; by allowing the internet to compel us to attend to a constant stream of different, trivial things, we have become unable to focus on the monolithically important thing that it is.

1

A Sudden Acceleration

> In 2010 you could say something like "if it's free, you are the product" and feel smart for a full year. These days you need to say something like that every few hours.
>
> —DON HUGHES[1]

> "Internet, c'est vraiment du Leibniz sans Dieu" ["The internet: it really is Leibniz minus God"].
>
> —MICHEL SERRES[2]

Our Critical Moment

The earth has moved under our feet in just the past few decades. The largest industry in the world now is quite literally the attention-seeking industry. Just as in the nineteenth and twentieth centuries the global economy was dominated by natural-resource extraction, today the world's largest companies have grown as large as they have entirely on the promise of providing to their clients the attention, however fleeting, of their billions of users.

And these users are, at the same time, being used. One vivid and disconcerting term that has begun circulating in social media to describe anyone who spends time online is "data-cow." The role that users of "free" online platforms occupy might sometimes feel creative, or as if it has something in common with traditional work or leisure. But this role sometimes appears closer to that of a domesticated animal that is valuable only to the extent that it has its very self to give. We do not usually provide our bodily fluids, and are not usually asked to do so, though sites such as Ancestry.com do ask for saliva as part of their data-collecting efforts, and health bracelets and other such devices owned by Apple and Amazon are increasingly discovering ways to monitor a number of our vital fluid levels. But even if we are not giving our fluids, we are giving something that has proven more valuable to the new economy than milk ever was in the system of industrial agriculture: information about who we are, what we do, what we think, what we fear. Some of us continue to have old-fashioned careers in the twenty-first century—we are doctors, professors, lawyers, and truck drivers. Yet the main economy is now driven not by what we do, but by the information extracted from us, not by our labor in any established sense, but by our data. This is a revolution at least as massive as the agricultural and industrial revolutions that preceded it. Whatever else happens, it is safe to say that for the rest of all of our lifetimes, we will only be living out the initial turbulence of this entry into a new historical epoch.

This then is the *first thing* that is truly new about the present era: a new sort of exploitation, in which human beings are not only exploited in the use of their labor for extraction of natural resources; rather, their lives are *themselves* the resource, and they are exploited in its extraction.

The great engine that is fed upon countless little nibbles of individual human attention, and that must constantly solicit such attention if it is to get fed, runs much more effectively, and is much better able to indulge its voracious appetite, when it appeals to human passion than to human reason, when it entices our first-order desire for dopamine-fueled gratification, than when it invites us to cultivate moral character or pursue long-term goals of betterment of self or world. This gives rise to what we may describe as a general "crisis of attention." Parents complain of the difficulty of limiting their children's screen time; the pharmaceutical industry develops new drugs and new avenues of profit in the fight against attention deficit disorders like ADHD; start-ups sell special brain-scanning goggles that shock students back into focus when their attention begins to flag; people of all ages complain that they are no longer able to read a book from cover to cover or even to watch a movie without slipping away to Google some half-remembered trivia about one of its players. The crisis is real, and many-tentacled.

Just as the overproduction of material goods is best understood in terms of its ecological consequences, the new crisis of attention is best understood in similarly ecological terms: as a crisis affecting a particular kind of natural being in a particular sort of informational landscape, one replete with human-made powers and dangers. As Yves Citton has sharply discerned,[3] with the rise of the internet, global human society has passed into a stage of overproduction of cultural goods (we had already long been living with the overproduction of material goods, however unequally distributed). In these new circumstances, by what means our eyes become locked on this fragment of human intention rather than that one is now among the most pressing matters in both politics and economics, yet understanding it requires us to pay close attention to how the human mind cognizes its surroundings and navigates its way through the world. Both cognitive science and phenomenology thus appear germane in new ways to basic questions of politics and economics.

This then is the *second new problem* of the internet era: the way in which the emerging extractive economy threatens our ability to use our mental faculty of attention in a way that is conducive to human thriving.

Both the first and second problems are aggravated significantly with the rise of the mobile internet, and what Citton astutely labels "affective condensation." Most of our passions and frustrations, personal bonds and enmities, responsibilities and addictions, are now concentrated into our digital screens, along with our mundane work and daily errands, our bill-paying and our income tax spreadsheets. It is not just that we have a device that is capable of doing several things, but that this device has largely swallowed up many of the things we used to do and transformed these things into various instances of that device's universal imposition of itself: utility has crossed over into compulsoriness. Our networked computers and mobile devices are not, or are no longer, analogous to Swiss army knives that include a few blades, scissors, a file, a small magnifying glass. That may have been the goal of some technologists as they sought ways to absorb the CD player, the book, the telephone, the camera, the daily calendar, the clock, etc., into a single universal device. But all this absorption has brought us to a transformation not just in the nature of our tool use, but in the contours of social reality. As the editors of *n +1 Magazine* presciently stated it as early as 2007: "The work machine is also a porn machine; the porn machine is also a work machine." This remains true even if you abjure pornography, and even if you are unemployed.

Whatever your habits and your duties, your public responsibilities and secret desires, they are all concentrated as never before into a single device, a filter, and a portal for the conduct of nearly every kind of human life today.[4]

This then is the *third feature* of our current reality that constitutes a genuine break with the past: the condensation of so much of our lives into a single device, the passage of nearly all that we do through a single technological portal. This consolidation, of course, helps and intensifies the first two novelties of our era that we identified, namely, the extraction of attention from human subjects as a sort of natural resource, and the critical challenge this new extractive economy poses to our mental faculty of attention.

It gets worse still. In Vladimir Nabokov's 1957 novel *Pnin*, the titular character is a lost and hapless White Russian emigré teaching Slavic literature at a university with a striking resemblance to Cornell University. He boards in the home of an American family, the matron of which, Joan, enjoys sitting at the kitchen table with him as she reads the fat Sunday newspaper. When she asks why he will not take a section and read along with her, he replies, sadly: "You know I do not understand what is advertisement and what is not advertisement."[5]

If such uncertainty was possible in the 1950s for a relatively underacculturated immigrant, today Professor Pnin's statement seems positively prophetic of a general condition to which even the most savvy navigators of our cultural landscape are prone. If we all find it difficult to distinguish between advertisement and not-advertisement, this is in part because, today, all is advertisement. Or, to put this somewhat more cautiously, there is no part of our most important technology products and services that is kept cordoned off as a safe space from the commercial interests of the companies that own them. The relatively small Twitter, and the much larger Facebook and Google, make their profits almost exclusively from advertising. While Amazon and Apple have different business models, obtaining much of their profit from the sale of goods, a significant part of their success is based on their ability to insinuate their logos, and more subtly their general aesthetic and one might even say their spirit, into the lifeblood of society.

Of course, the traditional American newspaper also made its profits from advertising, and by the 1950s major corporations such as Coca-Cola had imposed themselves not just as products, but as ideas and even as weapons in the Cold War battle for global cultural hegemony. But the scale was much smaller, and it remained fairly easy to opt out, as Pnin politely does with the newspaper. Moroever, while the *Ithaca Journal* attempted to draw and hold the attention of the reader in a way that would maximize exposure of its commercial sponsors to readers, it was in the end only ink on paper, a technology incapable of *reading those readers* in turn: incapable, that is, of compiling and exploiting even approximate engagement metrics. The new advertisement landscape by contrast is one that functions bidirectionally, monitoring potential customers' behavior, attentional habits, and inclinations, and developing numerous technological prods and traps that together make it nearly impossible to decide to exit this commercial nexus.

All of this is part of the extractive economy of attention we have already identified. But perhaps the greatest change over the past decades has been that individual readers or consumers are themselves now pushed and pressured to operate online according to the same commercial logic as the companies whose products they are using. In a basically pleasant conversation I had on a recent podcast, my host used a phrase that would become indelibly seared in my memory. Prefacing an observation about different styles of social-media use, the affable young man began: "Whether you are a brand, or an individual presenting as a brand." The listing of these two possibilities has the superficial character of a distinction, but its real effect is one of elision. Another podcast on which I was invited to appear, at around the same time, sent me an automated message beforehand advising me, during my appearance, to "Make your Brand look and sound its best." Both of these appearances were for promotion of my previous book, *Irrationality: A History of the Dark Side of Reason*, in which, significantly, I had at least something to say about the irrationality of human beings conceptualizing themselves as brands. But there is simply no other choice. You must use the internet in order to do anything at all, including writing and promoting books, and the more you use the internet, the more your individuality warps into a brand, and your subjectivity transforms into an algorithmically plottable vector of activity. Under these circumstances, one wants to say: "I do not even understand *of myself* what is advertisement and what is not advertisement."

This then is the *fourth genuine novelty of the present era*: in the rise of an economy focused on extracting information from human beings, these human beings are increasingly perceived and understood as sets of data points; and eventually it is inevitable that this perception cycles back and becomes the self-perception of human subjects, so that those individuals will thrive most, or believe themselves to thrive most, in this new system who are able convincingly to present themselves not as subjects at all, but as attention-grabbing sets of data points.

The earth, again, has shifted under our feet. We are the targets of a global corporate resource-extraction effort on a scale the world has never before seen. This effort harms us in numerous ways, not least by compromising our ability to use our faculty of attention in ways conducive to thriving. This compromise is felt most sharply in the condensation into a single device, no more than a few inches wide and a few inches long, of nearly everything that matters to us, often including even our interpersonal relations (or, as on Tinder or Grindr, hopeful attempts at them). For many, the only available adaptation to this new landscape is to transform our human identity into a sort of imitation of the decidedly non-human forces that sustain the internet, to trade a personality for an algorithmically plottable profile, in effect, to imitate a bot.

Paying Attention

Bots can do many things. They can monitor, track, harrass, impress with their ability to generate natural-seeming sentences, and even make jokes. But in the end they are like the cardboard-cow cutouts of a Potemkin village, as they are not themselves capable of conjuring that precious resource the new economy is intent on extracting: to wit, attention.

Attention is special among mental faculties for a number of reasons. Perhaps first among these is that it is not only a mental faculty, but also, irreducibly, a moral state. The moral aspect of attention is conveyed in familiar situations, such as the plea that one might extend to a loved one: "Pay attention to me!" It is also evident in the word's most common verbal form, "to attend," which can mean either to show up where one is expected, or to serve someone in a devoted fashion, not to mention the French cognate *attendre*, "to wait," a term that is not overtly moral, but that in common usage often implies a sense of duty.

Perhaps because of its partially moral quality, attention has been of at most secondary interest in the history of modern European philosophy, where many schools of thought since

René Descartes have tended to treat philosophy of mind and epistemology as if these could be largely cordoned off from practical philosophy. In modern theory of knowledge, it is much more common to find philosophers speaking in terms of "consciousness of" a given object, rather than "attention to" that object. These notions of course overlap, but what they have in common should not obscure their differences. While "consciousness" is semantically linked to "conscience" and thus also blends into the moral domain (indeed in French there is only one word for both notions), as it is used in modern philosophy it often concerns only the cognitive state of a subject, not the will or the dimension of the self in virtue of which that subject becomes a locus of praise and blame.

The notion of attention by contrast highlights the selection process involved in taking certain objects *rather than others* as the objects of one's consciousness, even though there are other objects in the field of one's perception to choose from if one wishes. Although theories differ, and although this has been a topic of much debate in philosophy at least since William James,[6] this latter faculty, perception, appears to precede and to lie deeper than attention. Perception is the cognitive state that one might say is "without a filter." This is what made it possible for G. W. Leibniz to speak of the *petites perceptions* by which bare monads, which is to say non-spiritual and non-rational mental beings, represent the world to themselves, even though they lack consciousness. Leibniz calls "apperception," in turn, the conscious awareness of a perception.

But attention is neither perception nor apperception. One may attend without being consciously aware that this is what one is doing. In fact, when we are intensely concentrating on something, say, performing a musical solo, it is most likely that we are attending to the sequence of chords involved, while doggedly not attending to the fact that this is what we are doing. To become aware of what one is doing in the middle of a difficult musical or athletic performance is to become self-conscious, to lapse out of one's "zone," and open oneself up to error. Likewise, in more mundane exercises of attention, we find they evade categorization in terms of the poles of perception and apperception. I look out at a grove, and several different species of tree present themselves to my sight. Among them is a eucalyptus tree, but I have not yet noticed this: I am currently focused on the oak. Nor am I thinking about this focus, let alone about myself as the one who is focusing. In such a state we may say that I am perceiving the eucalyptus, but not attending to it. The screening out of the eucalyptus in favor of the oak, moreover, is in the end a process involving the will. It is a choice, even if I am not always aware of making this choice.

It may be helpful at this point to summarize briefly the state of some of the recent literature in the philosophy of attention, conceived as a subdiscipline within the philosophy of mind. I take Carolyn Dicey Jennings's 2020 book, *The Attending Mind*,[7] to be both a fine representative of the state of the field of philosophy of attention as well as a means to mark out an interesting new direction for its further development. Much previous scholarship, at least if we confine ourselves to the Anglophone sphere and to the most recent decades, has been principally concerned to give us an account of what attention is *like*; Dicey Jennings sets herself the very different task of explaining attention in relation to its *source*. For her, that source is the *subject* or *self*, which she understands as a source of change, and thus as bearing causal power. "We experience being a source of change," she writes, "when we make an effort to change our behaviour."[8]

Dicey Jennings acknowledges that this experience may well be illusory, and for this reason, she maintains, we should also draw on what she takes as secondary and tertiary evidence: namely, and respectively, the tools developed in experimental contexts to identify the so-called active and passive behavior of subjects, as well as the evidence from studies of the brain for neural states corresponding to a subject's reported effort. Thus, phenomenology, experimental psychology, and neuroscience, she argues, furnish three mutually corroborating types of evidence for a subject lying behind acts of attention.

Attention, on Dicey Jennings's most succinct definition, is an act of mental prioritization by a subject, essential for perception, but not for consciousness or action. The subject is the one who "pulls the bow,"[9] to use a metaphor that draws on the etymology of "attention" in the verb *tendo,* whose primary meaning derives from archery. In this account Dicey Jennings positions herself against what she takes to be the predominant theory of attention as "selection from limitation,"[10] which she identifies in philosophers from St. Augustine, who contrasts our human inability to processs everything at once with the omniscience of God, to Min-Shik Kim and Kyle Cave, for whom "at any given moment the visual system receives more information than it can fully process. Thus, some portion of the visual input must be selected and processed more carefully than the rest."[11] While arguing that attention reveals its own causal sources in a subject or self, Dicey Jennings aims to avoid what she calls the "homunculus fallacy"—for her the self evidenced by attention is not a Cartesian metaphysical subject, but is rather brain-based and emergent. In other words, hers is an account that remains faithful to naturalism, even if it revives an entity, the self, most familiar from non-naturalist theories.

One attractive alternative theory, pursued by Jonardon Ganeri in his 2017 book, *Attention, Not Self,*[12] revives a distinguished line of inquiry from classical Theravada Buddhist thought, most notably as represented by the fifth-century CE philosopher Buddhaghosa. For Ganeri, attention is an action, which may be conscious or unconscious. It is attention that explains our active involvement with the world, what he calls our "freedom from passivity." But he rejects what he calls "the Authorship view," one element of which is the causal theory of action according to which an event is an action just in case it is caused by a rationalizing intention, which in turn is the result of an agent's motivating beliefs and desires. For Ganeri, rather, according to the "Attentionalist" view, "being at the center of an organised arena of experience and action is a property not of a real but at best of a virtual entity, which as such cannot have any causal powers; so the self cannot be an agent."[13] This is Ganeri's redux of Buddhaghosa's elaboration of the Pali notion of *anattā*, equivalent to the Sanskrit *anātman*, or "no-self."

The "arena" of experience Ganeri describes is in turn his gloss on the Pali *citta* (the same in Sanskrit). This term might best be translated as "frame," and in Buddhaghosa's theory it is precisely attention that provides the frame of experience and agency, rather than attention being a "space of awareness distinct from and occupied by experiences *of* awareness and agency."[14] For Ganeri, the account of attention as "the ongoing structuring of experience and agency"[15] provides a key for the interpretation of numerous satellite notions, such as mindfulness—"a kind of rehearsing or retentive attention"[16] and empathy—"attention through embodied comportment to the feelings, commitments, and wishes of others."[17] This latter notion, like "attention," conveys an idea of moral investment, yet it is historically a late arrival on the scene: a retrofitting of Greek word roots for an English calque of a German neologism, *Einfühlung*, coined only in the nineteenth century and attributed variously to Hermann Lotze[18] and to Robert Vischer[19] (also found in earlier variant forms in J. G. Herder). Ganeri nonetheless feels justified in using the English term to render two Pali words often employed by Buddhaghosa: *karuṇā,* most commonly translated as "compassion," and *mettā,* or "kindness."

It will be worthwhile to dwell for a moment on the particularities and possible sources of the notion of empathy in nineteenth-century German aesthetic theory, rather than running it together with other partially overlapping notions from either classical India or contemporary moral psychology. Intriguingly, Robert Vischer's use of the term, developing some themes found in his father Friedrich Theodor Vischer's aesthetic theory,[20] describes not so much interpersonal or second-person encounters—at least not in our ordinary restrictive sense, but rather the *aesthetic* relation that a person can have with an object, a work of art, or the world around them. What consideration of the history of modern aesthetics thus suggests is that determination of whether a theory of attention must trace back to an attending subject may be the wrong question to pursue. One might do better to think of attention to *objects* as their elevation into an intersubjective relationship, however temporary.

We may or may not wish to presume a subject behind our own acts of attention (Dicey Jennings does, Ganeri does not); and almost nobody today would want to presume a subject lying behind the phenomenal presentations of objects (though in the past Leibniz did just this). But in the end our differing metaphysical commitments may not matter so much for a robust and compelling account of attention. Attention, whether between persons or toward a work of art or toward a mousepad or a tree, may fruitfully be thought of not as including empathy as one of its instances, but rather with empathy constituting its most exemplary instance, while all instances, whether involving other persons or "mere" objects, partake somewhat of the phenomenological character of the second-person encounter.

Such an analysis is familiar, fashionable even, in certain corners of the philosophy of technology, and of what is known as "science studies."[21] It owes much to Martin Heidegger and the analysis of the notion of *thing* as one that derives originally from a social and political usage, whose most accurate definition would be something like "object of interest or solicitude."[22] Thus, for example, ancient Germanic societies were governed by a *Ding*, literally a "thing," but in context something more like a "council" or "assembly." The same notion remains today vestigially in, for example, the Icelandic word for "parliament," *Alþingi*, which can be glossed as "the All-Thing," and this term is in turn parallel, morpheme for morpheme, to the Latin *res publica* ("republic"), which we might literally translate as "the public thing." In this light, to relate to a given portion of the external world in its *Dinglichkeit* ("thingliness") is inevitably in some sense to allow it to occupy a place within a social scheme of values and interests that we ordinarily do not imagine it to have "on its own."

In the exercise of such concern it is not that we are literally experiencing, to speak with Ganeri, an empathetic "comportment to the feelings, commitments, and wishes" of inkpots or mousepads, but only that the special power of attention as a mental faculty is to generate the experience of subjecthood, and to do so bidirectionally. Bare observation without attention—if such a thing is possible—leaves the observer unchanged, or merely records the presence of the object; attention opens up the attender to the object, so that it may as it were go to work on the individual, bringing that person into a changed state and thus exercising a power we ordinarily restrict to agent-others worthy of a second-person stance. Or, as the tenth-century CE

Shaiva philosopher Abhinavagupta—whom Ganeri places in conversation with Buddhaghosa—puts it: "Whatever manifests itself has the shape of an 'I.'"

I hope this somewhat rapid detour, hereby concluded, through the philosophy of attention in both contemporary philosophy of mind and classical Indian philosophy will have been sufficient to give some force to the following claim: one way of understanding our present technocultural moment is that the intermediation of digital avatars and algorithmic tools makes the manifestation of other I's, and the second-person encounter that would result from this manifestation, difficult to achieve. The internet is an impediment to the cultivation of attention.

This does not suggest that we do not ever attribute subjecthood in our online experiences, but only that these attributions—weak, fleeting, and misdirected—cannot furnish the depths of intersubjective experience that sustained attention, attention that is not solicited from several directions, may be expected to deliver. In our current technological reality (and its fictional representations), we have ample illustrations of the ease with which human minds are in fact prepared to attribute subjecthood to automated systems, often in full knowledge that this is what they are. Consider for example the 2013 Spike Jonze film *Her*, in which a man falls in love with a Siri-like voice-command search-engine technology. Or, at a more mundane level, consider the experiments using AI-based humanoid and and animaloid dolls in Japanese rest homes, programmed to say affectionate and reassuring things to the elderly people with whom they've been partnered.[23] Researchers have found that the use of such dolls has significant effects in curing or reducing depression. It is not that the residents have been tricked, but only that having the dolls there was enough for them to encounter a simulation of empathy that allowed them to experience its positive effects.

So far, we have considered instances in which subjecthood is projected where it is not to be found. But likely far more common in the phenomenology of internet usage, particularly of social media, is the reverse: the presumption that one is interacting with an algorithm when in fact one is, or may be, dealing with another human being. It has even become a common insult in some corners of the internet to denounce a person with whom you disagree as "a bot." The implication is that their opinion is so crude that it may as well have been automatically generated. But to level such a denunciation is also a means of evading any threat of proper intersubjectivity, any obligation to display moral commitment to the other.

The relationship between self and bot, however, may be evolving. Over the past few years a new practice has been emerging of setting up bot accounts that mirror a person's real accounts, perhaps eventually replacing them. The bot versions of themselves automatically generate novel sentences that are built up from earlier real sentences they have written. In time, we might expect that the moral character, or perhaps merely the coolness or popularity, of people will be judged not on anything they actually say, but on the character of the bots that shadow them. In 2040, we might check in on how our digital personae are doing in the same way we periodically check our stocks: just as we now monitor our passive income, we may someday soon do the same for our passive selves; perhaps these selves, too, will be assigned a social-credit rating, whether called by that or some other name, to which we can appeal, or that we will seek to conceal, when we are maneuvering for elite university entry, or a new job or a first date. Already today, tech commentators such as Balaji Srinavasan are promoting the idea of pseudonymous employment, where we gain jobs and remuneration for work that is done entirely through our online avatars, without divulging our "real" identity to our (likely only short-term) employers.[24]

In much online interaction, it is plain that the question whether the source of what is happening on our screen is a real human or a bot is quite beside the point. It is well known that political polarization and the spread of conspiracy theories in recent years has been greatly exacerbated by the incentives built in to social media, where a subtle, nuanced, hesitant observation is likely to get you hundreds of times fewer likes and retweets than a bold declaration of partisanship. It is likewise well known that many who get pulled into the dynamics of like-seeking do not, to say the least, experience their online activity as a Turing test. That is, automated engagement with their partisan posts will do just as well as human engagement; both trigger the dopamine-reward system equally well, and even if one might have some lingering doubt about the ontological status of the being or the code behind the like one has just received, it is preferable, or rather more conducive to pleasure, to bracket that doubt as well as possible.

The account I have begun to sketch out here is in tension, if not in direct conflict, with other common analyses of the threats that the internet poses to attention. One popular representation of the challenges we face today launders the bit of folk wisdom according to which there is poverty in riches. As James Williams argues in an influential 2018 book, *Stand Out of Our Light: Freedom and Resistance in the Attention Economy*: "[I]nformation abundance produces attention scarcity."[25] But this seems to get wrong both the nature of the mental faculty in question and the condition of the milieu in which it finds itself. Even if we acknowledge we are operating within the domain of metaphor, it seems a mistake to speak of attention as subject to scarcity. For unlike fossil fuels, and much like love, attention appears to be the sort of thing that is capable of automatically replenishing itself from moment to moment. Even—or perhaps especially—if we are alone in a purely natural environment, there are infinitely many things that might hold our attention, and if we are not in control of this faculty we can find ourselves flitting noncommittally from one seashell or pine cone to the next, in a way that precludes any real appreciation of either. The same is true of artificial environments in past eras: as we will discuss later, in the seventeenth century the author of *The Anatomy of Melancholy*, Robert Burton, found reason to complain of all the trivial distractions that pull him away from serious thought, with only books, maps, letters, and slow-arriving broadsides doing the pulling.

The great difference, however, is that the books in Burton's library, like the features of the natural world, invite *sustained* rather than flitting or fleeting attention, as do the features of the natural world, and there are long and venerable traditions that have sprung up to better enable people to cultivate this attention. The internet, by contrast, seems to be structured so as positively to forbid such cultivation. This might be simply a condition of infancy. The great masterpieces of early cinema after all, such as the Lumière Brothers's 1896 *L'Arroseur arrosé* [*The Waterer Watered*], lasted no more than a minute. Roughly seventy years later we had Andy Warhol's *Empire* of 1964, more than eight hours long, which, if not a masterpiece, at least is an illustration that, among other things, the cinematic art may sometimes solicit us to undertake marathon exercises of attention that would have been inconceivable in the early years of the technology that made the new art possible.

Books seem to be intrinsically constituted so as to demand sustained attention from those who engage with them. It is difficult to imagine taking on Marcel Proust's multivolume magnum opus, for example, without what we might call a deep "attentional commitment" (whether we follow through on that commitment is another matter), one that has a moral dimension if for no other reason than that we know that the sheer amount of time we will have to devote to reading *In Search of Lost Time* is time taken away from our fellow human beings, and even from other forms of care of ourselves that we might have pursued. It is a big deal to commit to reading.

I myself have spent far more time over the past year scrolling through Twitter than I have spent reading literature, but I do not recall consciously making any such attentional commitment. It is in part for this reason that my scrolling strikes me as a moral failure on my part, and at the same time a moral wrong against me on the part of those who contrived to reduce me to this condition for profit. In part, a person is able to be sucked in, to a deep commitment that does not appear as such, because social media conceal the true manner of their working. Because our attention is being constantly solicited from multiple sources, and it is seldom held by any of these sources for more than a few seconds or minutes (which would be against the economic interests of the owners of these sources), we are able to pretend that the larger patterns of our engagement are something less than a profound existential transformation, a transition into a new form of life.

In the future there may of course emerge forms of engagement with internet-mediated arts, and other expressions of the human creative and intellectual drive, that will require the sort of conscious commitment we now make when we begin reading a weighty novel. And of course books as well have often—and often rightly—been condemned as trivial distractions, with no justifiable claim on our attention. Indeed the century and a half following the printing revolution in Europe, as Ann Blair has compellingly shown, was a period experienced by many who lived through it as one of "information overload." Even though the total bytes of information that had been stored in books, broadsides, papyruses, and so on from the beginning of human history up to the early modern period was extremely small, almost vanishingly small, in comparison with digital information storage today, it was quite enough to make many people feel as if there was already too much of it.[26]

If some who lived in the period felt that Gutenberg had ushered in a crisis of attention, thus anticipating in important respects our current moment, it is helpful to consider the forms of attention that were in the course of being lost as new technologies of information storage and transmission were on the rise in the sixteenth and seventeenth centuries. Thanks in particular to the influential scholarship of Frances A. Yates,[27] it is well known that throughout the Middle Ages tremendous value was placed on what was described as an "art of memory," an elaborate system of training that enabled students to master entire bodies of knowledge through the use of mnemonic techniques. Memory, we might say, is attention to something absent, so the loss of an art of memory could well be seen as one of the factors in the early modern crisis of attention. It is not that the technology of writing was unavailable to medieval scholars, but only that, for the most part, until the modern period *true* knowledge of an object of study involved internalizing that object by committing it to memory.

With the rise of printing, modern Europeans began their slow but certain shift to an understanding of "rote memorization" as something only backwards pedagogues and cruel school marms insist upon. But what we tend to ignore in the wake of this shift is that without memorization, without commitment to memory of a body of knowledge by means of an art of memory, all we have left is a distal relationship to that body of knowledge, where it is retrievable from storage (whether in a book or a hard drive or in the "cloud") should we need it, but which we cannot really be said to *know* until we retrieve it. I myself have come to feel as if the entire contents of Wikipedia is "mine," that I "know" it, that it is in some sense a prosthetic component of my own mind. But this feeling also involves a betrayal or abandonment of older dispositions to learning and of now antiquated norms of knowing. There is no progress without loss, of course, and it may be that the loss perpetually, as by an iron law, balances out the gain.

Nor is Europe alone in this loss of the art of memory. In classical India, notably, reading and writing were often perceived as a mere crutch for those who lacked the strength of memory to internalize a given work.[28] Some canonical "books" of ancient Indian literature did not exist at all as physical texts for several centuries after they were "written." Thus for example in the fourth century BCE, the great Sanskrit grammarian Pāṇini composed his masterpiece, the *Aṣṭādhyāyī*, by dictating different portions of it to selected students, who would then dictate their portions to their students in turn. This system remained in place for several centuries, until someone lost resolve and had to write it down. Pāṇini would have been disappointed in this latter-day scrivener; for him language is not language if it does not have the breath of life in it. The Indian and European cases are not exactly alike, but in each of them we see a transition away from memorization and toward external storage of knowledge, and with this transition we see the loss of a profound and venerable cultural practice.

The feeling of loss of something profound and venerable, when we see that today young people spend hours on end scrolling through Instagram and do not seem to even consider the idea of picking up a book and reading it from cover to cover, echoes in important ways these earlier historical instances of loss. Our most recent loss is already one stage removed from the understanding of knowledge as memorization and internalization, which already began to slip away with the rise of the printing press. Reading *In Search of Lost Time*, for example, involves the retrieval of externalized information, in contrast with, say, the recitation from memory of the *Aṣṭādhyāyī*. Proust put down on paper a century ago the content of his rich internal life, enabling us to get it back at will. One could in principle learn Proust's oeuvre by heart, but few people consider this a reasonable thing to attempt. Julien Sorel, the protagonist of Stendhal's 1830 novel *The Red and the Black*, is noteworthy for having memorized the Bible, and here we may have an example of a novelist reckoning with what has been lost in the shift from the art of memory to the practice of reading. In any case to read, again, involves at least a commitment, undertaken in the knowledge that one will be transformed internally through the act of reading: a transformation that may well be as profound as what one might expect to happen, cognitively and indeed emotionally, when a disciple of Pāṇini shapes his own mind into a storage device.

Yet another dimension of attention worth noting is that the thing to which one attends typically reveals itself in new ways as a result of this attention. If someone says that they are conscious or aware of an apple on the table, it is generally safe to say that that person is not in the process of discovering new things about that apple. But if someone tells you that they have gone to the museum and attended to a work of art (this language sounds precious and rare in today's English, yet we are still able to understand it[29]), you are thereby made to know that they have undertaken a concrete exercise, which like all exercise has a different character at the end than at the beginning. If you stare at a rock formation long enough, you will see that it is no longer just a rock formation, but becomes, perhaps, an incitement to meditate on geological time or natural order, or perhaps it becomes an occasion to discern imagined figures in its surface. And memory, as attention to something absent, also changes in character as the absent thing recedes into the past.

In the contemporary world there is little social value placed on such meditative exercises. Meditation has been largely coopted in the name of the distortion known as "mindfulness": a concept custom-made for the internet era, whose real purpose is maximization of conscious attunement to some goal, typically a professional or profit-oriented goal, lacking any of the openness to radical transformation of the self and deconstruction of one's goals that meditation, non-cynically understood, makes possible. To properly attend to something, unlike being mindful of it, is to relinquish control of the meaning it holds for you, to allow it the potential to become something else, something unfamiliar within the context of your prior range of references and expectations.

Today the true exercise of attention is nearly as unwelcome in "the real world," moreover, as it is on the internet. In an art museum, for example, you will likely be ushered along, after only a minute or two, by a security guard or a crowd of people hastily taking photographs for Instagram, and this is probably not nearly enough time for the work to begin its transformations under your sustained gaze. The temporality of the museum, one fears, is being distorted and accelerated by a cognitive disposition that is shaped by the internet and that creeps out from there.

The most sustained focus that the internet invites is perhaps that of coders reviewing their work. But this is, precisely, *work*, and its purpose is to weed out mistakes, not to see an object in a new way or to allow one's mind to move down unexpected paths. In this respect, coding does not involve attention as we are defining and explicating it here. One would perhaps do better to call it "concentration." Attention, as we have been considering it, cannot involve laborious correction of mistakes in its object, but rather is a matter of moral and cognitive investment in an object or text *as it is*, which in turn opens up the possibility of transcending it, or understanding abstract principles beyond it, or even gaining new forms of self-knowledge.

Thus, I skim the first volume of *In Search of Lost Time* and it seems as if the narrator has his priorities all wrong, that he is self-absorbed and long-winded. I start to imagine ways I might correct the work, as if it were a string of code and I were a programmer looking to make it "run" more optimally. Then I read on and I begin to feel myself committing morally to the narrator, which means, among other things, opening myself up to receiving the world as he apprehends it. As this process continues, and I sink deeper into the work, I find that that world is not at all as it first appeared to me. It is not that I now think the narrator is "good" or "correct" or "praiseworthy," whereas at the beginning the opposite appellations applied. It is that the moral commitment that I have taken on toward him, the commitment of attention, has infused his world into mine. Neither the novel nor the reader is any longer the same.

It is an understatement to say that the internet has not been good so far at providing opportunities for the exercise of such attention. But, again, there does not seem to be anything about the technology itself that would explain this failure. Rather, it is much more likely that the failure is to be explained by consideration of the economic model that drives online engagement: a model that, again, maximizes solicitations upon a user's attention and ensures that the attention is never focused in one place for long. It is not that "[i]nformation abundance produces attention scarcity," as Williams argues.[30] There is in fact an infinite amount of information in every tree, in every pond, yet typically we do not think of the solicitations these natural entities make on our attention as unwanted distractions or as impediments to the cultivation of this faculty. Rather, information abundance produces attention scarcity when the information is being processed through an engine that is explicitly designed to prod the would-be attender ever onward from one monetizable object to the next. Under such conditions, while the extraction of attention remains the basis of the new internet economy, the cultivation of *individual* attention amounts to a form of hard-won resistance against this economy.

If we agree to consider that *transformative moral commitment*—whether to another person or to a work of art or perhaps also to nature—is not only an instance *of*, but a clavis *for* understanding attention, then we may in turn consider the following possible account of the present "crisis of attention": that it results not from the overabundance in the digital age of possible foci for this mental faculty, as James Williams suggests, but from the excessively narrow channeling of our cognitive and emotional investment down pathways that are structurally guaranteed to limit or prevent personal transformation. Whether this is a temporary limitation of our digital technologies, deriving from the economic model that governs them, or whether it is here to stay, is a question it is likely still too soon to answer.

Gadget Being

"I do not pretend to know what is moderation, or where the line is between legitimate and illegitimate gadgets."[31] These words were written by the environmentalist Aldo Leopold in the 1940s, concerning what he saw as a worrisome proliferation in his era of new technological devices intended to improve the bounty of the modern hunter.

We might at first glance suppose that hunting is an activity the philosophy of which would have little overlap with the philosophy of art or literature or education, and that would not be thrust into crisis by the advent of new technologies in the same way these other domains are. But just like these other domains, hunting requires attention—even, one might venture, a form of reading. And like these other domains, hunting is an activity that emerges from necessity, but that also contains within it rich material for reflection on profound questions. When Leopold says he does not know which gadgets are legitimate and which are illegitimate (use of the term "gadget" skyrocketed from the mid-1930s through the 1950s, then plummeted, only to begin to make a partial comeback since the turn of the millennium), he is identifying a general problem of our condition.

Our technologies are born as enhancements of our species-specific activities. To devise an artificial duck call or decoy is not to cheat at duck hunting, but is rather very much a part of duck hunting itself. Yet to become so reliant on such accessories as to lose the ability to attend to ducks, in their behavior, in their nature, is to leave off from the primordial experience of hunting and to begin doing something else altogether. At what point a gadget ceases to enhance and begins rather to distort or pervert, an activity, will differ for different people, and so it is always futile to attempt to distinguish between good and bad gadgets as such. Presumably, a better criterion for judgment is offered by a consideration of what the gadget is doing for the person who uses it, and how the gadget shapes the understanding of the activity in which it is implicated.

If we agree with Leopold, the true end of duck hunting cannot simply be the accumulation of dead ducks, and so success in duck hunting cannot be determined by a simple consideration of the day's yield, the "metrics" of the activity. Although the hunter does not have the option of throwing back his prey, as the fisher does, in hunting as perhaps more sharply in fishing it is widely recognized that the activity itself is one of its own ends (as for example is paradigmatically the case in activities such as strolling or daydreaming, in contrast with, say, house-building). The success of any activity of this sort cannot be measured in "outcomes," and in this respect, as Izaak Walton well understood of fishing in his remarkable book of 1653, *The Compleat Angler*, these activities share something important with philosophy.

It should be clear enough that the contemplative and destinationless character of hunting and fishing, which makes them at least partially philosophical in nature, is a character also shared by many sorts of conversation, by many forms of writing, and by most forms of reading. In light of this, one way of understanding the crisis into which the internet has thrust us is that it has deprived us of this character that reading, writing, and communicating naturally share with philosophy (and indeed with hunting) by aggressively metricizing them. One might put this another way and say that whatever is metricized or gamified, whatever keeps itself centered before the user's mind with the promise of accumulating more points of some currency (likes, faves, followers, up-votes, not to mention also—with the rise of "meme stocks" and cryptocurrency exchanges—real money), has the power to harness and hold the user's concentration, but not their attention.

In its current form it is as if we have had imposed on ourselves a gadget that does nothing more than count the number of ducks we have killed, broadcast that number to the entire world, rank us with that number alongside all other duck hunters, and invite all of them, as well as whatever interlopers feel so inclined, to praise, criticize, or mock us for our ranking. Such a device would obviously not be an enhancement of the activity of duck hunting as such (it dawns on me as I write this that there most likely *is* such a device, and that duck hunters have been pulled to it just as surely as, say, scholars have been pulled to Google Scholar). Social media, similarly, are not enhancements of communication. Academia.edu and other websites that operate according to social-media principles, bombarding anxious young scholars with spam-like messages pretending to inform them that "a senior scholar" has just downloaded their paper, that their work is being clicked in the top 1 percent on the site, and otherwise nudging them to continue to "build their brand": these platforms are not enhancements of scholarship, but rather scholarship-themed games, bearing the same relationship to our work as thinkers that, say, *Grand Theft Auto* bears to driving stolen cars.

We have by now arrived dangerously close to a thicket of metaphors: hunting is a sort of reading, fishing is a sort of philosophy, and so on. The term "gadget" in particular seems especially susceptible to metaphorizing. No particular tool is "truly" a gadget, but is always a tool of this or that sort. We use this word when we want to discuss tools as such, and to do so in a mildly mocking and diminutive way. Such a comic tint is there wherever, in particular, gadgethood is projected back on the users of gadgets, when so to speak the waterer gets watered, and human beings begin to be conceptualized on the same ontological model as the tools they have created. When Jaron Lanier warns that "you are not a gadget," as he does in the title of a well-received book,[32] he is attempting to win back something, let us call it the irreducible human essence, that has long been perceived to be under threat, from Descartes's thought experiment of the automaton that walks outside his window under a hat and cloak, to the cultural representation of the comical, but also uncanny and horrifying, 1980s cartoon figure of Inspector Gadget, also under a hat and cloak, laying bare his mechanical inner workings in order to fight crime.

More recently still, we have not only philosophical thought experiments and amusing popular representations, but also a new anthropological model that has filtered out from the internet into the broader culture as an implicit understanding of who we are as human beings and how we work. In this new understanding, the comedy lies undetected by those who adopt it. It is common, now, to read on the internet accounts of human action that model it on artificial systems and that have no other resources for conceiving human motivation than those borrowed from programming, even when what is at issue is human moral failure.

To invoke one striking example, while also avoiding a needless account of the particular issue that occasioned it, a recent controversy was triggered by the behavior of a well-known comic-book artist accused by numerous younger women, most of them fans of his work and aspiring comic-book authors, of sexually inappropriate behavior, emotional manipulation, and what is known as "grooming," in which a person in a position of power shows personal interest and attention to a person of less power, only in the aim of lowering the latter's defenses for the purposes of exploitation. A website was set up for his proclaimed victims to share their testimonials. On this site, the author's grooming behavior is described as "rel[ying] on subtle techniques that leverage 'compulsion loops,' which are well-established in scientific literature and video gaming, and are commonly utilized by modern businesses to achieve addiction, AKA 'user retention.' Examples include daily quests in games, getting a higher reward (more 'XP,' etc.) for the first game of a day, more 'karma' for the first post of a day on a message board, etc. The main driver is a regular daily dopamine boost sustained over time."[33] At issue here is the moral conduct of a person who in another era would have been accused of lechery, of moral transgression, of playing the cad, or of sinning. Here the accusation against the comic-book author, however, eschews inherited moral categories and blames him, effectively, for instantiating the same features we also know from our use of social media. The author has had programmed into him, we are told, the same addictive hooks for which we rightly criticize Facebook. He stands accused of user retention.

Such reduction of others to a sort of program is the flip side of what we have already identified as "presenting as a brand," and both are expressions of the more general problem of what we may call "algorithm creep": the tendency to see an ever broader portion of the world, and even to see ourselves, on the model of the algorithms that run our new technologies. As will be discussed later, in recent decades even certain questions in speculative metaphysics (notably the question whether our world is a "simulation") have been substantially influenced by the spread of a manner of thinking about reality that is borrowed from the world of video-game design. This habit of thinking extends to everything from fundamental questions about the nature of the external world, to questions about the structure of human society, from electoral politics to interpersonal relations. To "game the attention economy" is to develop a strategy for "winning" attention from others within a system of formal constraints where the "points" are measured out in clicks, likes, favorites, retweets, and so on: the quantified units of attention. But a person may be gamified by this new economy in turn, either taking their own behavior variously as a sort of well-running or glitch-laden program, or falling victim to denunciations from others for one's own glitches.

When we criticize Facebook for its manipulative ploys for user retention, we are in fact criticizing the people who designed Facebook, and the people who signed off on this design for reasons of greed. But who, in turn, are we really criticizing when we criticize the comic-book author? One plausible answer is that we are criticizing Facebook, and all the other internet platforms and video games that promote such a system of motivations as the one of which the author is being accused. On another interpretation, the author has consciously programmed himself, acting in a way that simulates the systems of motivation he has studied on the internet. Either way, we are in uncharted territory, far outside the realm of traditional praise and blame, as now the normative evaluation of human action cannot proceed without "running it through the machines," without reference to the behavior of artificial systems that are themselves insusceptible to praise and blame.

This transformation is shaping the way we understand not only interpersonal relationships, but also political movements. According to the video-game designer Adrian Hon, the QAnon conspiracy theory, whose supporters became a vigorous and multitudinous force backing Donald Trump in the later phase of his presidency, might best be understood as an "ARG" or "alternate reality game."[34] Such a game is not played on a console; instead its strategies and prizes are spread across the internet, built into apps, inserted into newspaper advertisements and even into real-world interpersonal relations. Yet it is conceived and executed on the model of a traditional console-based game, and designers and users of traditional video games are generally the people who are most invested in the construction and navigation of such alternate realities. According to Hon, QAnon is "a uniquely 21st century conspiracy theory," in that it rewards its supporters for "going down rabbit holes" and making connections between disparate clues in order ultimately to discern an underlying unity. Arriving at such unity is the standard desideratum of conspiracy theories in general, and now it is also a sort of grail or ultimate reward after which a new species of gamers-cum-conspiracy theorists are questing.

It has been reported, similarly, that the Uber corporation has "adopted techniques typically used in video games to more effectively manipulate their drivers."[35] Such examples of what we might call "existential gamification"—where the game model encroaches on domains of human life, such as politics and work, from which it was previously thought to be quite distinct—will no doubt continue to multiply in the coming years. Corporations will manipulate people into behaving as if they were video-game players, and these same people will adopt an understanding of their own freely chosen pursuits in life as if life were a video game. "Game" is a notoriously difficult term to define; Ludwig Wittgenstein enjoyed using it as an example of something for which no necessary and sufficient conditions could possibly be proferred, containing everything from peek-a-boo to chess to "war games." What is certain, though, is that in the spread of gamification to so many dimensions of social life, the connection of the idea of gaming to "play," in the old sense evoked by Friedrich Schiller of the imaginative exercise of freedom, is growing ever more tenuous.

As already mentioned above, G. W. Leibniz got the underlying philosophical problem of artificial intelligence right: when we outsource our thinking to machines, we are not bringing new subjectivities into existence, new conscious beings like ourselves. We do not have the power to do that, nor do we have the power even remotely to understand what would be involved in such an act of creation. Rather, outsourcing our decision procedures is really only getting machines to run a simulacrum of thought, one that has everything our own thinking has, except perhaps for the subjectivity, the presence of a conscious mind behind the thinking.

Leibniz was wrong, however, about the usefulness of this outsourcing for human life. Or at least he was unable to anticipate the down side. Leibniz had anticipated that the more drudging and uninteresting operations of the mind might be outsourced to machines: they'll do the math and the analysis of arguments, so that we might "think big," contemplating ideas and synthesizing the results the machines give us into new and original arguments. What has in fact happened, it often appears, is that these drudging and uninteresting operations, as they are fed back to us, are being taken for the highest form of intellectual work possible, and are being used to model, and in turn to constrain, the way we understand our own minds and our own wills. Leibniz thought we should parcel out to machines that part of our intellectual work that they can do, so that we might concentrate on that part that they cannot do. Now, instead, one may fear, we have perilously neglected this remainder, and while the machines were originally modeled after us, we now take what they can do as the ultimate form of intellectual work, and we emulate it, modeling ourselves after them.

Let us illustrate this last point with a particularly vivid example. There was a time, before the ubiquitous application of algorithms to our social life, when it was considered salutary, when it was deemed central to a full and meaningful life, to actively cultivate one's aesthetic sensibilities, for example, one's musical taste. The "You may also like" function of music-delivery platforms such as Spotify has largely obviated the need for such active pursuit, and now a person who happens to start exploring music from a given song in a particular genre will typically be guided along to other songs categorized as similar to the first one based only on criteria that AI is capable of "understanding."[36] This means that a listener who happens to hear, say, Björk's 1999 rendition of "Gloomy Sunday" (also known as "The Hungarian Suicide Song") will likely only be guided toward other Björk songs, rather than back into the magnificent history of this work for which she is in fact only serving as a vessel, through Lydia Lunch's 1977 rendition, Mel Tormé's of 1958, Billie Holiday's of 1941, and all the way back to the original 1933 version of Rezső Seress, "Szomorú vasárnap." The same thing happens if one first hears Marilyn Manson's version of "I Put a Spell on You": the algorithms do not think to introduce us from there to the versions of Nina Simone or Screamin' Jay Hawkins or the Birthday Party, but instead keep the listener trapped in that inane, market-defined cage known as musical "genre."

In this latter instance, the listener will continue to be exposed to mostly mediocre instances of *fin-de-millénaire* industrial goth. Not that there's anything wrong with this genre, or any other, but it is a fundamental misrepresentation of the life of a pop song to map it only synchronically next to its genre peers, rather than to see it diachronically, as something that, if it has any staying power at all, will weave across different genres, always transforming while still maintaining the same essential character. This is the beauty of popular music, and it is what gives the cultivation of aesthetic appreciation for it a distinct character, distinct especially from appreciation of classical music, where, however much subtle variation there may be from conductor to conductor, from performance to performance, nevertheless fidelity to the original intent of the composer is valued much more highly. So to be pushed along from Marilyn Manson to some other contemporary genre peer or near neighbor, rather than back to Nina Simone, is to be prevented from cultivating an aesthetic sensibility appropriate to the art form one has begun to explore.

Things gets worse still, much worse. In 2018, Spotify struck up a partnership with Ancestry, one of the premier companies offering DNA analysis that purports to reveal a customer's ethnoracial background. Spotify users were to have the option of integrating their DNA test results into their listener profile, which in turn was to direct the algorithm to bring songs to the playlist roughly reflecting the percentages of the listener's ethnic background. Thus, if the DNA test revealed that a person had 10 percent Irish origins, every tenth song might belong to the thoroughly commercialized genre known as "Celtic folk." An advertisement produced for this new service asked, "If you could listen to your DNA, what would it sound like?" The answer, it turned out, at least for the partially Irish among us, was that it sounds like Enya and *Riverdance*.

This is, to put it mildly, a gross betrayal of the meaning of both music and ancestry. If someone had to take a DNA test to learn that she was partially Irish in the first place, then there is simply no meaningful sense in which she has any more truly inborn receptivity to Celtic folk music than anyone else in the world. That is just not how music and culture work. In this respect the partnership both builds on and contributes to the crude essentialist ethos predominant in the present moment, which takes individuals to belong to cultures absolutely and essentially, and which discourages movement across the always fuzzy borders of cultures as a violation of the imperative to "stay in one's lane," and as the new crime of "cultural appropration." This ethos is a pop-cultural refraction of the disastrous lurch into nationalism and isolationism throughout the world over the past decade, a historical process that has thrust such bigots and trolls as Donald Trump, Narendra Modi, and Viktor Orbán into power.

We are not, yet, accustomed to seeing these different trends—the corporate opportunism of Ancestry and Spotify; the sinister right-wing populism of the aforementioned leaders; and the identitarian campaigns for cultural purity driven mostly by young self-styled "progressives" on social media—as inflections of the same broad historical phenomenon. But perhaps their commonality may become clearer when we consider all of them as symptoms of an underlying and much vaster historical shift: the shift to ubiquitous algorithmic management of society, which lends advantage to the expression of opinions unambigous enough (i.e., dogmatic or extremist enough) for AI to detect their meaning and to process them accordingly, and which also removes from the individual subject any deep existential imperative or moral duty to cultivate self-understanding, instead allowing the sort of vectors of identity that even AI can pick up and process to substitute for any real idea of who an individual is or might yet hope to be.

The Tragicomedy of the Private Commons

The revolution in attention-extractive profit-seeking extends well beyond our small screens and permeates all aspects of contemporary life, even for those remaining few people who do not use social media, smartphones, or apps. Hollywood movies are now commonly test-marketed using third-party companies that track the eye motions of the test audiences as they are watching the screen, in order to determine by quantitative means what sort of images and motions succeed best in capturing a viewer's attention. As the test viewers are watching the screen, the screen is literally watching them back, learning from them in a heuristic process that ensures that subsequent movies, and the final edits of the same movie, will be ideally fitted to the first-level cognitive-emotional cravings of viewers—which is to say, at the same time, that these movies will be absolutely unable to challenge the viewer in any way, to cause the viewer to grow in heart or mind by exposure to something new and unexpected.

This is, in a very crude sense, giving audiences what they want, and in that same crude sense it enables the entertainment industry to pretend that it is allowing viewers to exercise their "freedom of choice." But of course this is a perversion of most any idea of freedom that has ever been taken seriously by any philosopher. Few would want to say that the heroin addict is more free when in possession of a dose of heroin and about to hit up, than when circumstances prevent him from getting what he wants and he finds himself desperately jonesing and unhappy. A standard way of explaining the apparent contradiction in saying that a person who actively goes out and gets what she wants is nevertheless not exercising her freedom, is to maintain that sometimes there are irrational first-order desires that are in conflict with rational second-order ones. Since the nineteenth century, many have been captivated by the idea that mass culture—whether religion, as for Karl Marx, or popular music and movies, as for Theodor Adorno—is in some metaphorical sense an "opium." More recently China's delivery of TikTok to the West, with the political and cultural upheavals it has triggered, has recently been described as "revenge for the Opium Wars," sending back, after a century and a half, a new sort of addictive drug that also threatens to exacerbate geopolitical instability.[37]

With TikTok and similar online platforms, the comparison of mass entertainment to opium may now be passing from metaphor into literalism. The science of addiction is revealing that the brain's reward system works in largely the same way whether the hit it is receiving comes as true opium or as a like on Facebook. The like is probably administered by another human being—probably, but not necessarily—but its effect is largely the same, and is meant by the algorithm's designers to be the same, as a well-placed video of a funny cat or of wicked police brutality; the same as the targeted ads for hypertension remedies that start to hit a person after evidence of a hypochondriac fit is detected in the person's search history; as the coupons that arrive by mail from Target on the basis of past purchases and data-mined predictions of future ones; and as the Marvel superhero spectacle that has been custom-made from the movements of test audiences' eyeballs. All of these forces, whether inducing a good feeling or a bad one, are, like opium, making us less free and less capable of achieving true human thriving.

Although this system is relatively recent, at least in its all-encompassing drive for monetization of every aspect of our lives, a few things are by now very clear. One is that, although it has by now engulfed aspects of our lives that do not take place online, the primary motor by which this system has spread and is perpetuated is the internet. Another is that although the internet is the primary motor for the spread of this new system, it is very likely that whatever new sanctuaries we might yet hope to build, where the rapacious logic of this new system does not have any purchase, will *also* come through the internet.

Remarkably, then, the internet is simultaneously our greatest affliction and our greatest hope; the present situation is intolerable, but there is also no going back. Against Lanier's plea to "close your social media accounts,"[38] we must recognize that such a gesture is futile in a world that has been revolutionized in all of its domains by the influence of the social-media model.

The only solution to the problems the internet has left us with is to shape the internet, or at least certain corners of it, into something else. This is not to advocate using "the master's tools" to break free. The internet, though it is oppressing everyone, never belonged to any particular master. It has had quite a bit of government funding and planning, but it is hard to say which of such contributions were absolutely crucial for its development, and it seems likely that it would have emerged, along a somewhat different timeline, out of the cumulative effect of small actions of organic tinkering by individual users. The internet is, as we will argue in the next chapter, an *excrescence* of the human species, and it is thus, by rights, ours.

I have already accused social media, in the introduction, of failing to do what those most hopeful about the potential of this new technology anticipated it would do, namely, that it might facilitate deliberative democracy, that it might serve as the public space in which ideas are debated and conflicts resolved or overcome, in which citizens work together to advance toward better and more just solutions to shared problems. What we have in fact obtained in place of this is a farcical imitation of deliberation, in which algorithms are designed by the companies that provide the platforms for discussion in order to maximize engagement, a purpose that is self-evidently at odds with the goal of conflict resolution or consensus-building. Social media are in this respect engines of perpetual disagreement, which sharpen opposing views into stark dichotomies and preclude the possibility of either exploring partial common ground or finding agreement in a dialectical fashion in some higher-order synthesis of what at the first order appear as contradictory positions.

Anyone who has even dabbled in social media can experience this limitation directly. What keeps people coming back therefore cannot be that they have failed to notice that the platform is useless for achieving their end, the end of persuading others or, alternatively, coming to a new position oneself as a result of the persuasiveness of the others. Rather, one comes back for the fight itself, perhaps (as in my own user experience) with feelings of both regret and giddy anticipation, the same murky mixture of motivations that drive one forward in any addictive behavior at all. And yet we also know that social media are, increasingly, the only game in town when it comes to public discourse, when it comes to "getting a message out" or "making yourself heard." This leaves many people with a civic education built on realities from the previous century wondering what to do: people who feel the need to "speak up" but are deeply averse to unproductive dialogue, especially when it is being controlled by corporate interests that do not *want* it to be productive.

Civic education has long been based on the idea that public engagement is *good*, that it is part of our duty as citizens to read the news, to write letters to editors, to talk to people in our social circles about important issues of the day. But the news is now mostly a current within the ocean of social media. On a typical day in 2021, more than one of the articles from the salaried op-ed writers at the *New York Times* will often consist largely in their summaries of what they have read on Twitter. For those of us who have already read the same thing, the *Times* cannot help but strike us, at this point, as functioning as a sort of filter that normies prefer to place on social media to soften, or perhaps to launder, their own engagement with it. They put the *Times* filter on Twitter in the same way one might use a photo filter app such as Hipstamatic on one's selfies, to avoid having to see all the blemishes and imperfections in their vivid and unaltered directness. If newspapers no longer exist except as a filter option for social media, then of course the old form of engagement known as a "letter to the editor" is at this point only a quaint exercise, and everyone knows that the more potent form of response is to post a snide dismissal of an article or of the columnist who wrote it, an ad hominem denunciation of the columnist's character, perhaps tagging that columnist's employer and insinuating or stating directly that the terms of the columnist's employment should be reconsidered.

It is difficult to know, yet, what a site for the exchange of ideas would look like that did not rely on algorithms designed in the interest of profit-seeking. But such a thing is of course technically possible, and what we are seeing for now is a sort of perversion of collective deliberation, as people, some of them with sincere good will, seek to use what is essentially a privately owned point-scoring video game as if it were the public sphere. This has led to an absurd predicament for the good-faith actors and a luscious opportunity for the bad-faith ones. As the pseudonymous Twitter user known only as "Alice from Queens" sharply observes, Twitter is the place where "socialists show contempt for hierarchy, meritocracy and neoliberal competition by competing for status in a game designed by a Silicon Valley overlord."[39]

Nor is this a situation that could have been fully anticipated by the theory of the "tragedy of the commons," at least if we understand tragedy in the proper sense of a misfortune that befalls a protagonist as a result of some particular blind spot, like Oedipus, who did not recognize his own mother. The tragedy of the commons, as ordinarily understood, is the universally undesirable result of the collective actions of individual parties acting out of self-interest within a shared-resource system. But social media are not a shared resource in the traditional sense. Twitter users are not "using up" space or bandwidth or electricity in the way that cattle use up the grass on public land designated for grazing.

Users do compete for limited attention, and a successful poster typically finds advantage by learning to game the algorithms with insincere or artificial phrasings that they seem to pick up more readily. But the limits on available attention are not set in the same way as the limits of available edible grass are. Rather, attention is artificially limited, channeled, and redirected for the sake of profit. Users moreover are often not acting in a self-interested way—or not purely. They are not like hungry cattle, but rather spend much of their time "signal-boosting" perfectly worthy altruistic purposes (whether this altruism is diluted by a share of virtue-signaling is irrelevant to the present point: a cow when it grazes is neither signal-boosting nor virtue-signaling, just eating). The tragedy, then, is very different from the tragedy of the commons: it is that individuals, even those operating for non-self-interested reasons, end up contributing to a situation that is universally undesirable, because they are operating in a system or field (to continue the grazing analogy) that only simulates a true commons, but in fact operates to the detriment of individual users no matter how they are behaving, whether in self-interest or altruism, whether seeking to destroy or to build.

If the medium is the overarching problem that shadows any attempt to engage with any other problem within that medium, then how is one to engage at all? Trying to advance a political or social cause through social-media activity can feel somewhat like aiming a firehose toward a tidal wave. The hose and the water it emits were designed to solve a problem, but when the water comes together with the vastly greater body of the same element that is approaching, it only disappears into the rising wave of catastrophe. Every Tweet, no matter what its content, no matter whether true or false, righteous or trolling, adds to the tidal wave that is currently crashing over our public space and

2

The Ecology of the Internet

Signals

The internet is still not what you think it is.

For one thing, it is not nearly as newfangled as the previous chapter made it appear. It does not represent a radical rupture with everything that came before, either in human history or in the vastly longer history of nature that precedes the first appearance of our species. It is, rather, only the most recent permutation of a complex of behaviors that is as deeply rooted in who we are as a species as anything else we do: our storytelling, our fashions, our friendships; our evolution as beings that inhabit a universe dense with symbols.

In order to convince you of this, it will help to zoom out for a while, far from the realm of human-made devices, away from the world of human beings altogether, to gain a suitably distanced and lucid view of the natural world that hosts us and everything we do. It will help, that is, to seek to understand the internet in its broad ecological context, against the background of the long history of life on earth.

Consider the elephant's stomp: a small seismic event, sending its signature vibration to kin over a distance of kilometers. Or consider the clicks of a sperm whale, which, it is now thought, can sometimes be heard by familiars on the other side of the world. And it is not just sound that facilitates animal telecommunication. Many or perhaps most signals sent between members of the same species pass not through sonic vibrations, but through chemicals. Female emperor moths emit pheromones that can be detected by males more than fifteen kilometers away, which, correcting for size, is a distance comparable to the one traversed by even the most resonant sperm whale's click. Nor is there any reason to draw a boundary between animals and other living beings. Numerous plant species, among them tomatoes, lima beans, sagebrush, and tobacco, use airborne rhizobacteria to send chemical information to their conspecifics across significant distances, which in turn triggers defense-related gene expression and other changes in the growth and development of the recipient.[1] Throughout the living world, telecommunication is more likely the norm than the exception.

At this point some might protest that "telecommunication" is being used here in an equivocal way, as for example when we say of both an irate cyclist at an intersection and of our computer when it freezes up with its cursed rainbow disc spinning, that they are "angry" at us. Some might object that even if for the sake of argument it is conceded that the sperm whales and the elephants are sending out signals that may be processed *as* information, that is, as a symbolic encoding of propositional content that is then decoded by a conscious subject, the same surely may not be said of lima beans.

Let us grant, if only to avoid unnecessary complications, that lima beans are not conscious. We may still ask why, when telecommunication in both conscious and unconscious life forms evidently involves the same principles and mechanisms, we should be so quick to assume that telecommunication in our own species must be the product of consciousness, rather than being an ancient system that arose in the same way as lima bean signaling, and only belatedly began to allow our human consciousness to ride along with it. The former assumption seems to get things exactly the wrong way around: telecommunication networks have been around for hundreds of millions of years. Isn't it possible that the most recent outgrowths of our own species-specific telecommunicative activity—most notably, the internet, but also such systems as telegraphy and telephony, which we take to be extreme departures from the previous course of human history—are in fact something more like an outgrowth latent from the beginning in what we have always done, an ecologically unsurprising and predictable expression of something that was already there?[2]

And could it be, correlatively, that the internet is not best seen as a lifeless artifact, contraption, gadget, or mere tool, but as a living system, or as a natural product of the activity of a living system? If we wish to convince ourselves that this suggestion is not mere poetic rhapsodizing, but something that is grounded in a sort of truth about both technology and living systems, it might help to consider the long history of attempts to imagine telecommunication technologies through the model of animal bodies and vital forces.

"All things conspire"

Human telecommunication requires not just knowledge of how to build devices to capture signals, but also some understanding of the nature of the medium through which those signals move. One common cosmological theory in antiquity took the universe itself to be a sort of living body, and thus imagined that physically distant parts of the physical world are in constant feedback relations with one another, where any change in one region is echoed or mirrored in any other, just as the pain of a rock landing on the extremity of my foot is felt not only in my foot, but also in my physically somewhat distant head. The universe was thus a "cybernetic" system, in the sense described by Norbert Wiener in the mid-twentieth century, whom we will discuss in some detail later. Like the animal and the machine for Wiener, the universe as a whole for many ancient theorists was characterized by a circular causality or signal-looping.

The causal interconnectedness of all parts of an animal body was well captured in the Hippocratic (or more likely pseudo-Hippocratic) motto, *Sympnoia pantōn*, which may be translated variously as "The conspiration of all things," or, in a somewhat more literal but also exactly equivalent rendering of the verb *conspire*: "The breathing-together of all things." Conspiracy, when we break down the elements of the word, is nothing other than synchronized or harmonized breathing. The Hippocratics were physicians, and they understood this motto to encompass the interconnectedness of the parts of the body, the way in which my lungs filling up with air is also a replenishing of the life of my toes, and fingers, and the top of my head; the way in which my foot's pain is also my head's pain; or the way in which an illness of the kidneys may give rise to symptoms and morbidities in other parts of the body. Later philosophers, notably in the Stoic tradition, extended this account of physiology to the world as a whole. Thus the Stoic philosopher Marcus Aurelius, invoking the metaphor of weaving, implores us to think of the universe as a single living being, observing "how intertwined in the fabric is the thread and how closely woven the web."[3]

If the web of all things is so closely woven, then nature itself, independently of the tools we develop to channel it or tap into it, already possesses the potential for near-instantaneous transmission of a signal from one place to another. It is just this sort of transmission that our wireless communication today realizes. But we did not need the "proof of concept" that has finally arrived in only the most recent decades in order to feel the force of the conviction that it must, somehow, exist.

Those ancient authors who recognized the possibility of telecommunication generally understood that while the natural medium through which signals are to be sent may preexist humanity, we are nonetheless going to have to rely on our own technological ingenuity to tap into and exploit that medium. The devices envisioned by these authors were often rather simple, and even in their own era were perfectly familiar and mundane. In the first-century CE fantasy novel, *A True History*, the Greek-language author Lucian of Samosata imagines a trip to the moon, where he discovers a "mighty great glass lying upon the top of a pit of no great depth, whereinto, if any man descend, he shall hear everything that is spoken upon the earth."[4] This is a principle of simple amplification, whose proof of concept is already present whenever a person enters a seaside grotto or a cave that causes voices to echo.

To some extent, telecommunication just is amplification: simply to speak to a person in a normal voice is already to telecommunicate, even if at naturally audible distances we have learned to be unimpressed by this most of the time. But with a glass or a saucer or ear trumpet, the ordinary qualities of sound waves are magnified, and the possibility for total global surveillance of all conversations from a satellite (in Lucian's example a natural one) of our planet becomes thinkable.

Often, in early attempts to appropriate natural forces for telecommunicative ends, it was not a matter of amplifying known powers of nature, but of manipulating nature in new ways to draw out hidden or merely suspected powers. Thus, in the seventeenth century, the English natural philosopher Kenelm Digby sought to harness the power of "the weapon salve," an old treatment for soldiers injured in battle that worked through manipulation not of the victim's wound, but of the weapon that wounded him. Digby saw that if an alteration in the state of a dagger could bring about a simultaneous change in the body of the person it had previously stabbed, then in principle such a force could be seized for transmission of messages across long distances as well. An anonymous pamphlet drawing on Digby's work was published in London in 1688, proposing that a stabbed dog might be put on a ship moving across the Atlantic Ocean, while the dagger that stabbed the poor beast might be manipulated every day at the same time, causing the dog to howl in pain. And in this way the precise time at the place of departure could be determined, and from the angle of the sun and other measurements the lines of longitude could be demarcated, and a major navigational hurdle overcome.[5] "[W]e might at Sea with great Ease and Pleasure know when the Sun was upon the Meridian at London," the author notes, while anticipating the obvious humane objection: "Fye! says one, or other, you would not sure put a Dog to the misery of having always a Wound about him to serve you, would you?"[6]

But the dream of torturing animals in order to unleash their innate telecommunicative potentials is one that would not go away easily. In the middle of the nineteenth century a French anarchist and con man by the name of Jules Allix managed to convince at least a handful of Parisians that he had invented a "snail telegraph," that is, a device that would communicate with another paired device at a great distance thanks to the power of what Allix called "escargotic commotion."[7] The idea was simple, if completely fabricated. Based on the widely popular theory of animal magnetism proposed by Franz Mesmer at the end of the eighteenth century, Allix claimed that snails are particularly well suited to communicate by a magnetism-like force through the ambient medium. Once two snails have copulated with one another, he maintained, they are forever bound to each other by this force, and any change brought about in one of them immediately brings about a corresponding change in the other: an action at a distance of the sort that the mechanical physics ascendant since the seventeenth century had sought to banish, but never fully succeeded in banishing.

After all, what is Isaac Newton's theory of gravity but action at a distance? It is precisely the evident non-mechanical implications of gravitational theory that caused Leibniz to reject Newton's theory, to insist that no body may attract any other body from afar, since to move by attraction, rather than by the pulling or pushing of subvisible corpuscles, is to have the sort of internal soul-like power that modern physics was intent on denying to bodies. Thus Leibniz was wrong about the particulars—gravity does exist, or so our best theories tell us for now—but he chose the wrong answer for the right reason. Allix's reasoning, in turn, was that we know as a *fact*, observable in everyday experience, that gravity, not to mention magnetism, is real, and there must therefore be something incomplete about the physical theory that had been intent on denying action at a distance. And it could very well be, Allix suggests, that other types of such action are still waiting to be discovered in nature and perhaps waiting to be harnessed for technological applications.

Against this theoretical background, Allix takes, or pretends to take, two snails that have previously copulated, and he places each of them in its own small slot on its own device, each of which corresponds to the same letter of the French alphabet. Then the two devices are removed from one another, and messages are sent from the one to the other by successively

FIGURE 1. Honoré Daumier, "Les escargots non sympathiques." From *Le Charivari*, September 25, 1869. The snails are spelling out the word "PROGRESS."

manipulating the snails in the appropriate slots in order to spell out French words. In a feigned demonstration given in Paris in 1850, Allix receives the message: *LUMIÈRE DIVINE* (*DIVINE LIGHT*) from a correspondent purportedly in America (it is not explained how all the snails were transported so far after copulation, without any of them perishing in the voyage).

Allix predicts that at some point it will be possible to make pocket-sized devices using particularly tiny species of snails, and that we will then be able to send messages throughout the day—"texts," you might call them—to our friends and family as we go about the city. He envisions being able to receive newspapers from the whole world on these devices, and to follow the deliberations of parliament. When Allix is exposed as a grifter, he absconds from Paris, having already taken the money of his gullible investors. He reappears a few years later on the isle of Jersey, leading spirit-summoning séances in the presence of, among others, a skeptical Victor Hugo.

The story of Jules Allix reminds us that a rigorous historian of science may learn just as much from the fakes and frauds as from the genuine article: even when someone is lying, they are nonetheless doing the important work of imagining future possibilities.

Nature's Technique

Allix's device, as he envisions it, is in a sense a species of wifi. The would-be inventor knew that the earliest telegraphy had required two conductive wires—one for the signal to go out and another for it to return. But, as Allix explains, after experiments in Paris beginning in 1845, it was proven that the earth itself can function as a conductive medium and can thus take on the role of one of the two wires. His project, then, is to allow nature to replace both of the wires, and to allow the incoming and outgoing signals to be conducted between the two devices through a medium that preexists both devices as well as the human desire to telecommunicate. In this minimal sense, the sperm whale's clicks, the elephant's vibrations, the lima bean plant's rhizobacterial emissions, and indeed Lucian's listening disc, are all varieties of "wifi" too, sending a signal through a preexisting "ether" to a spatially distant fellow member of their kind (and also, sometimes, to competitors and to prey of different kinds).

It was just as common from antiquity through the modern period to envision nature not as pervaded by an ether, but so to speak as a wired or connected network, that is, as a true and proper web: a system of hidden filaments or threads that bind all things. Such a system is instanced paradigmatically in what may be thought of as the original web, the one woven by the spider, presumed in many cultures to be the first inspiration for all human textile weaving (a technology to which we will return below).

The spider's web may be properly—meaning not only metaphorically—considered as the locus of its extended cognition.[8] An arachnid's nerves do not extend into the filaments it spreads out from its body, but the animal is evolved to apprehend vibrations in these filaments as a fundamental dimension of its sensory experience. The spider's sensation is not "enhanced" by the vibrations it receives from the web, any more than my hearing is enhanced by the presence of a cochlea in my inner ear. Perceiving through a web is simply what it is to perceive the world *as* a spider (or at least as a member of one of the many spider species that spend a good part of their lives in webs of their own spinning).

We ordinarily imagine that our own webs of wires are enhancements, and not intrinsic to what it is to perceive as a human, to what it is to *be* a human, since they did not emerge together with the human species, but are only a much more recent addition to the repertoire of the species. The web of a spider is a species-specific and species-defining feature of the spider, while the internet, we usually suppose, is a superaddition to the human. We will return to this distinction, and the question of its tenability, later on. The important thing to register for now is that the spider's web is a web in at least some of the same respects that the World Wide Web is a web: it facilitates reports, to a cognizing or sentient being that occupies one of its nodes, about what is going on at other of its nodes.

Such webs may be found throughout nature. The natural webs that have lately enjoyed the most frequent comparisons to the internet are those we know from the vegetal world, whether a field of grass with its subterranean creeping rootstalks, or a grove of trees with its mycorhizal filaments connecting a vast underground network of roots, whose exchanges can now be tracked by a technique known as "quantum dot tagging."[9] Some of the earliest comparisons of such living systems to the internet were drawn from certain currents of twentieth-century French philosophy and adapted into communication theory in the early years of the internet. In the 1990s and early 2000s, the observation that certain features of human society, including human communication networks, might be "rhizomoidal" in character—that is, might have a structure resembling that of the subterranean networks of roots connecting the blades in a field of grass—was associated preeminently with the twentieth-century French philosopher Gilles Deleuze. In the influential 1980 work, *A Thousand Plateaus,* written together with Félix Guattari, Deleuze identified a number of characteristics of rhizomes, many of which seem equally to characterize the internet: a rhizome connects any point to any other point, a rhizome operates by spreading and offshoots rather than by reproduction, a rhizome has no center and no head, among others.[10]

As a result of his reflections on the rhizome, Deleuze, who died in 1995, is often regarded as an early visionary of the internet, whose vision was the more vividly confirmed, the more human beings came to rely on massive decentralized systems for their own daily communications. But this may be a case in which a philosopher's being right has actually impeded rather than enhanced understanding of what he got right. If the internet *is* rhizomoidal, then there should be nothing particularly Deleuzean about the observation that it is so. Nonetheless, the way philosophical fashions work often makes it difficult to see that, when a general theoretical claim is patently true no less than when it is patently false, a philosopher has no business trademarking it.

Quite independently of Deleuzean theory, in the past decade or so some plant scientists, along with their journalistic ancillaries,[11] have also come to appreciate the internet-like qualities of the underground systems of exchange, facilitated by bacteria and mycorhizal fungi, that are realized along the roots of trees. The "wood wide web," as journalists have called it, is a "complex and collaborative structure,"[12] in which trees enlist the assistance of numerous other life forms in order to maintain themselves and one another in good health, and also, it appears, in order to exchange vital information with one another at long distances.

We tend to suppose that whatever is species-specific or essential to a given biological kind cannot ineliminably involve another species, that what it is to be a panther or an oak ought to be something that could be spelled out without implicating fleas or moss in the description. But the tendency to think this way is mostly our inheritance of an inadequate and un-ecological folk-metaphysics. For example, scientists were so hesitant to see the fungus lining tree roots for what it was—namely, a life-preserving symbiont—that for a long time they took it to be a harmful parasite. In reality, symbiosis is common enough and central enough to the various species implicated in it that it is often impossible to understand what a species is in terms that bracket the existence of any other species. This is certainly true of the symbionts that make up the wood wide web.

The symbiotic relationship between fungi and plant roots is coevolved with the individual species involved in the relationship. If the relationship does not involve technology, in our usual understanding, it certainly does involve what Immanuel

Kant understood by the term "technique": the beings of nature, through their own internal capacity, making use of what is at hand, or at root, to bring about their proper ends. Thus, in the *Critique of the Power of Judgment* of 1790, Kant writes: "[W]e conceive of nature as technical through its own capacity; whereas if we did not ascribe such an agency to it, we would have to represent its causality as a blind mechanism."[13] The technique involved in symbiosis has also at times been compared to the process of animal domestication by human beings: for example, in the fungus/algae pairing that makes up the two-species life form known as lichen, the fungus is sometimes described as a sort of "algae farmer."[14] And if we agree with the commonplace that a domestic pig or goat is an "artificial" being, to the extent that it is nature transformed in the pursuit of human ends, why should we not also agree that the algae farmed by fungus or the fungus enlisted by the tree to pass chemical messages and nutrient packets along its roots (much as the internet is said to facilitate "packet switching"): why should we not agree that this technique is technology too? Or, conversely, and perhaps somewhat more palatably for those who do not wish to rush to collapse the divide between the natural and the artificial: why should we not see our own technology as natural technique?

Cetacean Clicking and Human Clicking; or, the Late-Adopter Problem

At least since Kant it has frequently been noted that living nature, or what we now call the biological world, presents a particular difficulty in our effort to distinguish between justified and unjustified carrying-over of explanations from one domain to another, and moreover that whatever justification there may be for doing so is not going to come from a deepened knowledge of empirical science.

When Kant proclaimed, also in the *Critique of the Power of Judgment*, that there will never be a "Newton for the blade of grass"—that is, that no one will account for the generation and growth of grass in terms of blind mechanical laws of nature in the way that Newton had managed to do a century earlier for the motions of the planets, the tides, cannonballs, and other objects of interest to mathematical physics—he was not simply reporting on the state of research in the life sciences, such that he would have to concede, if he miraculously came back to life, that the successive discoveries by Darwin, Newton, Mendel, Watson and Crick, and others, amounted to a sort of "collective Newton" for the blade of grass. Rather, Kant supposed, we will always be cognitively constrained, simply given the way our minds work, to apprehend biological systems in a way that includes, rightly or wrongly, the idea of an end-oriented design, even if we can never have any positive idea, or, as Kant would say, any determinate concept, of what the ends are or of who or what did the designing. In other words, we are constrained to cognize living beings and living systems in a way that involves an analogy to the things that we human beings design for our own ends—the clepsydras and ploughs, the smartphones and fiber-optic networks—even if we can never ultimately determine whether this analogy is only an unjustified carrying-over of explanations from a domain where they do belong into one where they do not.

Kant understood the problem as an intractable one, arising simply from the structure of human cognition. Yet this did not prevent subsequent generations from assuming dogmatic positions on one of the two possible sides of the debate concerning the boundary between the natural on the one hand and the artificial or cultural on the other. "Do male ducks rape female ducks?" is a question that sparked and sustained heated and ultimately futile debates in the late twentieth century. The so-called sociobiologists, led by E. O. Wilson, took it as obvious that they do, while their opponents, notably Stephen Jay Gould, insisted that rape is by definition a morally charged category of action and so also by definition a category that pertains only to the human sphere; that it is thus an unjustified anthropomorphization of ducks to attribute the capacity for such an action to them; and that moreover it is dangerous to do so, since to say that ducks rape is to naturalize rape and in turn to open up the possibility of viewing human rape as morally neutral. If rape is so widespread as to be found even among ducks, the worry went, then some might conclude that it is simply a natural feature of the range of human actions and that it is hopeless to try to eliminate it. And the sociobiologists would reply: perhaps, but just *look* at what that drake is doing, and how the female struggles to get away, and try to find a word that captures what you are seeing better than "rape."

The debate is, again, unresolved, for reasons that Kant could probably have anticipated. We can never fully know what it is like to be a duck, and so we cannot know whether what we are seeing in nature is a mere external appearance of what would be rape if it were occurring among humans, or whether it is truly, properly, duck rape. The same goes for ant cannibalism, for gay penguins, and so many other animal behaviors that some people would prefer to think of as distinctly human, either because they are so morally atrocious that extending them to other living beings risks normalizing them by naturalizing them, or because they are so valued that our own sense of our specialness among creatures requires us to see the appearance of these behaviors in other species as *mere* appearance, as simulation, counterfeit, or aping. And—to end our detour into much broader philosophical questions than the ones that concern us throughout this book—the same holds for the mycorhizal networks that connect groves of trees. Are these communication networks in the same sense as the internet is, or is the "wood wide web" only a metaphor?

It is not to be flippant or to give up too easily to say that the determination is ours to make, and that no further empirical inquiry will tell us whether such a comparison or assimilation taps into some real truth about the world. The choice *is* ours to make, though we would perhaps do better not to make a choice at all, but instead, with Kant, to entertain the evident similarity between the living system and the artifice with an appropriate critical suspension. Our minds will just keep returning to the analogy between nature and artifice, between organism and machine, between living system and network, and the fact that our minds are doing this says something about who we are and how we make sense of the world around us. So let us continue then, neither in a dogmatically literalist vein nor in an equally dogmatic opposite vein, and instead declare, in a critical spirit, what we in any case cannot help but notice: that, like a network of roots laced with fungal filaments, like a field of grass, the internet, too, is a growth, an outgrowth, an excrescence of the species-specific activity of Homo sapiens.

If this is what it is, or—to speak in a more critical spirit—if this is an idea we are constrained to entertain when we think about it, we are still confronted with what may be called the "late-adopter" problem. Sperm whales have been clicking for as long as there have been sperm whales, yet it is only in the past few hundred years, say, since Claude Chappe devised the flag semaphore or optical telegraph at the end of the eighteenth century, that we have been telecommunicating at all, and it is only in the past twenty years or so that telecommunication has become integrated into human lives in a sufficiently deep and widespread way for it to be taken for granted as simply "what one does," as inseparable from human flourishing. Significantly, it was in 2011 that Frank La Rue, special rapporteur for the United Nations, declared internet access a human right, since "the Internet has become a key means by which individuals can exercise their right to freedom and expression."[15] Consequently, in recent years the internet has begun to occupy a position in human life previously reserved for such familiar necessities as food, shelter, and clean water. Yet we know that there was a time—many of us can even remember a time—when there could have been no claim to such a right, while by contrast we have always needed nutrition and hydration. So again, if telecommunication is simply a natural consequence of our species-specific activity, it would seem to be so in a different way than the signaling of sperm whales is among their kind, for evidently we have not always telecommunicated. Cetacean clicking in the ocean, and human clicking on the internet, are categorically different.

Or are they? A great deal depends on how we define the thing in question. It is true that flag semaphore is the earliest realization of a system of communication across great distances that approximates simultaneity (if in turn by "great distances" we understand distances that are greater than a person could possibly see with the naked eye, thus eliminating, for example, smoke signals, which are truly simultaneous, or at least as fast as the speed of light). It took about one hour for a message to travel the 491 miles from Paris to Strasbourg, across semaphore towers positioned roughly six miles apart, their watchmen quickly noting the sequence of flags from the tower immediately to the west, and then transmitting it to the tower to the east. This is of course not instantaneous, but to be able to straddle in minutes a distance previously measured in days fed the hope of reducing the minutes to seconds, and those, eventually, to nanoseconds.

By the 1840s the optical telegraph lapsed into quick desuetude when electrical telegraphy entered the scene,[16] and within a few years of that we find Jules Allix holding forth—fraudulently, yet prophetically—about manipulating the forces of escargotic sympathy in order to eliminate the wires, and the delays, that came along with this new technology. Yet, as has already been made clear, it would be a mistake to suppose that telecommunication, as we understand it today, was "discovered" in the 1790s, or in the 1840s, or whichever terminus a quo we might choose. For one thing, Allix's escargotic sympathy is only a variation on Franz Mesmer's theory of animal magnetism (a term coined in 1779[17]), Digby's theory of the weapon salve (propounded in 1658,[18] and drawing on ideas already developed by Paracelsus in the early sixteenth century, and by many others before him), and many other hypothetical forces that connect the bodies of living beings in order to explain reciprocal, instantaneous cause-and-effect relations between them. Again, the possibility of near-instantaneous communication did not need to be demonstrated with metal wires and electrical contraptions in order to take hold as an idea, as an explanatory framework, and as a possible source of power for those human beings who knew how to seize it.

We may also wonder about the importance of instantaneity, or its semblance, as a component of the definition of telecommunication. Why is instantaneity so important? And do we even understand what we are talking about when we invoke it?

To a great extent it is telecommunication technology itself that determines what we understand by instantaneity, rather than new technologies helping us successively to draw closer to a pre-given and absolute ideal of instantaneity. We are familiar with the complaints among nineteenth-century intellectuals and critics that the trains on the newly created rail systems traveled too fast, even though their maximum speeds would trigger loud complaints of sluggishness if they were maintained on a typical European bullet train, or even on Amtrak, today. The perceived celerity of trains was enough to create the impression that the distances between locations were not just reduced but annihilated. Thus, in 1843, the German poet Heinrich Heine wrote that "Space is killed by the railways. I feel as if the mountains and forests of all countries are advancing on Paris."[19] What to us would be experienced as a leisurely jaunt appeared, to those who lived through the steam-power revolution of the early 1800s, to warp the very contours of time and space.

Nor, once again, does a technology—whether for rapid transit or telecommunication—need to exist in reality in order for the effect it triggers to be imagined. In the first century CE, Pliny the Elder writes of an Olympic runner who inadvertently discovered "time zones," or at least discovered the fact that it is simultaneously different times of day in different parts of the world, when he ran down a great Alpine slope to successively lower altitudes further to the west, and noticed that the day stayed with him as he went, just as it does with us when we fly from Europe to America. The French historian Serge Gruzinski intriguingly suggests that no question is more quintessentially modern than "What time is it where you are?," as it requires a grasp of the simultaneity of different times in different places around the world, and also requires the power to reach across the divide between these places in order to communicate with an interlocutor at a geographical remove.[20] In fact, though, Pliny was already capable of entertaining the theoretical question of the simultaneous reality of different times; what is different, for Heine, is that these different times now seem to be crashing into one and the same place, layering realities upon other realities that previously geography was able to hold apart.

The possibility of simultaneity—conceived as some form of immediate access across distinct realities—has always been thinkable, while new technologies make what was previously thinkable actual. And again, what is to count as simultaneity changes with successive revolutions in speed. While of course everyone involved in epistolary exchange in early modern Europe was aware of the lag between dispatch and receipt, a letter, from Bavaria to Paris, from Hannover to London, nonetheless seemed to speak directly across distances, and in the absence of telegraphs or text messages that could move ahead of the letter and circumvent the lag of its transit, it could not have seemed that the letter was in any regard temporally defective, any more than we find the light of a star inadequate when we are reminded that it was emitted eons before it arrived to earth.

Nor is the Republic of Letters as new as its participants imagined it to be. All the major ancient empires of Europe and Asia had a system for the transmission of letters by courier, though the Roman Empire is often held to have initiated the first mass-scale postal service. From the Roman *cursus publicus* through the medieval Mongolian *yam* to the early American Pony Express, the development of a system for delivering messages went hand in hand with the construction of passable roads across the extent of empires. Thus the "posts" from which mail systems derive their name are the stations positioned at suitable distances, at which a courier or a voyager could stop and let his horses rest, or, in the more developed and standardized systems, could stop and change his tired horses for fresh ones, both sets of which are owned by the same state monopoly.

But empires, whether ancient, medieval, or modern, were only institutionalizing practices that were already in place, and for which it would be futile to attempt to seek a moment of origin. While it is difficult (though not wholly unknown) to find personalized "messages" from among the material-cultural artifacts produced prior to the era of writing, we know from countless brooches, beads, carved bones, and trinkets of flint, amber, and lapis lazuli that long-distance networks of exchange reach as far back in human prehistory as we might wish to go, all the way back to what the archeologist Colin Renfrew calls the "speciation phase" of human evolution, when "biological and cultural evolution worked together to develop the human genome and the human species" roughly 60,000 years ago.[21] The best evidence from archeology and cognitive science suggests that complex material culture, symbolic art, language, and, importantly, long-distance trade, all emerge essentially in tandem as a suite of interrelated species-specific capacities. Even if Paleolithic people were not delivering letters across distances, it is certain that the seashell necklaces and other items we know they *were* delivering were perceived by their recipients to be packed with symbolic meaning, just like a letter, just like a sequence of emojis.

In other words, the "late-adopter" problem looks quite a bit different when subjected to real scrutiny. Human beings qua human beings are telecommunicators, just as sperm whales are click communicators and elephants are seismic communicators. Even in the vastly longest span of the history of human telecommunication, moreover, when messages could only travel as fast as a human being was able to run, there was likely no obstacle to conceiving of telecommuncation as simultaneous, as a collapsing of the rift between two separate realities so that people on both sides of the rift could share in the same reality.

"I see a vestige of man"

We have already spoken of "proof of concept," but this is only because the manner of speaking of the tech industry has to some extent rubbed off and caused us to state things less precisely than we should. It might be more correct to say "proof that the concept may be physically instantiated," for simply having the concept at all is its own proof: proof, namely, of the existence of the concept, which is an immaterial thing and so need not be physically instantiated in order to be real. Strictly speaking, then, telecommunication had its proof of concept long before it was physically realized by our devices.

In the early 1630s, for example, a French periodical reports on a certain Captain Vosterloch, who has visited an island in the South Seas, the inhabitants of which used a most noteworthy system for communicating across long distances. Beneficent nature, the article relates, rather than furnishing these people with the art of writing or the knowledge of science, had given them "certain sponges that retain the sound and the articulated voice . . . so that when they wish to ask something or to confer with someone at a distance, they simply speak next to one of these sponges, then they send them to their friends, who, having received them, press on them very gently in order to make the words that are inside of them come out."[22] This is not an internet-like device, but rather an anticipation of the cylinder, the tape recorder, or other sound-recording technologies that have existed in reality since the end of the nineteenth century. But again, the proof of concept did not need to wait for the physical realization of any of these devices. The proof of concept is already there in the 1630s, in the fable of Captain Vosterloch in the South Seas.

Much could be said about this remarkable text, beyond the obvious fact that the scene it relates is one of pure fantasy. But we will content ourselves with pointing out two things. First, it is significant that the author puts the islanders' practice in explicit relationship to European technologies in the broad sense, both the technique of writing and the scientific knowledge that might help them to develop artificial means of telecommunication. This strongly suggests an understanding of technology as, in the end, an approximation of what in the ideal case nature would give us for free, without need of any complicated transformations or rearrangements. Where nature furnishes people with sponges already equipped with a voice-recording function, the need to develop other technologies of communication at a distance, including writing, is correspondingly diminished.

Another element of this text worth highlighting concerns the relationship it suggests between orality and literacy. What made possible the idea of voice recording, long before there was any proof of concept (in the tech industry's imprecise sense) in the form of Thomas Edison's 1860 phonautograph device, is that writing itself could be understood in a certain respect as "voice recording," since it generates a *record* of what can also be said orally. If it were said orally, it would immediately evaporate, unless of course there were some other tehnological or natural system, such as the sponge, to capture it.

This is yet another example of the way in which even fantastical contraptions in the history of science and technology complicate our efforts to establish "firsts." The first voice recording was made in the mid-nineteenth century, unless we count writing, in which case we must move the date back several millennia earlier. But writing, in turn, is what enabled people to fantasize about other forms of voice recording, such as the recording sponge, which fantasies in turn may well have played a causal role in the search for the technology that would eventually become the phonautograph or the mp3. In short, there are no firsts, and the past, as St. Augustine said, is always pregnant with the future.

When the Christian philosopher made use of this latter image in the fourth century CE, he was, obviously, speaking metaphorically. The real nature of temporality does not involve biological generation; time cannot get pregnant. His use of this metaphor is grounded in a metaphysics of time that holds that what is real is simultaneous, while temporal flow is something produced by conscious experience. We do not have to share in this metaphysics in order to feel the same veridical force in the parallel metaphor concerning technology: prehistoric exchange networks, the Roman postal system, the Republic of Letters, the optical telegraph, were all pregnant with the internet.

Over the past decades there has been a significant shift in the way the public understands the idea of historical progress, and the way the pivotal revolutions in human history, notably the agricultural and the industrial revolutions, functioned as triggers of sudden and significant progress. In the mid-twentieth century, a standard tool for measuring a society's relative degree of progress was to measure its energy consumption. Thus in 1943 the anthropologist Leslie White argued that the basic law of cultural evolution is that "culture evolves as the amount of energy harnessed per capita per year is increased, or as the efficiency of the instrumental means of putting the energy to work is increased."[23] We sometimes revert to this way of thinking still today, when for example we look at a nighttime satellite image of the Korean peninsula, with the South blazing in artificial light and the North in near total darkness, and we imagine this to be a measure of the relative political enlightenment and social advancement of the two countries. But at the same time, and much more indicative of changes since White's day, we also praise countries like Denmark and Finland for reducing energy consumption and for having the maturity and sense to turn off unneeded lights at night. Indeed, we congratulate ourselves for reducing our own household energy consumption, and we take this to be a measure of our progress away from, rather than of our regression back to, a North Korean condition, or to what White would have called a primitive level of society.

This broad change in our standards of evaluating human societies results not only from ecological crisis—from the sharp awareness today, only just dawning in 1943, that we do not have unlimited energy to burn, at least not in its present forms and without serious consequences—but also from the innovative work of anthropologists, cognitive scientists, and environmental philosophers who questioned the idea that, as Marxist thinkers had it, the essence of man is to transform nature.

For one thing, it became increasingly clear in the second half of the twentieth century that the supposed revolutions in human history were largely historiographical fictions, or at least that they did not happen as we imagine them to have happened. The "human revolution" of the Paleolithic (when significant art and artifacts of material culture began to appear), the agricultural revolution of the Neolithic, the scientific revolution that inaugurates modernity, and the industrial revolution that announces modernity's maturity: for the most part, these successive revolutions were not experienced by those who lived them as sudden moments of illumination, as when an apple falls on the head or a cartoon lightbulb appears, and the benighted past, the old way of doing things, is instantly swept away. It is later history-writing that does the sweeping, whereas in the moment, historical continuity with the past, with the legacy of all who have gone before, remains inescapable. This is perhaps even more true the further back in the chain of purported revolutions we look. Steven Shapin wrote a book about the Scientific Revolution whose first sentence reads: "There was no such thing as the Scientific Revolution."[24] He nonetheless manages to describe at least some seventeenth-century natural philosophers who sincerely believed that they were making a radical break with the past, that they were "revolutionaries." By contrast, as James C. Scott has shown,[25] it is nothing short of preposterous to suppose that anyone who lived through the moment we now call the "agricultural revolution" saw the cultivation of plant crops as anything like a leap forward from an inferior level of social development characterized by foraging.

What is more, in recent years scholars have become attuned to the fact that human groups considered for a long time to be living in a "state of nature," to not have realized the essential human activity of transforming nature to any significant degree, in fact, from their own internal point of view, often envision the natural environment in which they live as a product of their own efforts, stewardship, and care. Thus, William Balée has vividly shown that for some Amazonian groups, the rain forest is not at all conceptualized as a "wilderness" in the sense in which westerners understand this notion, but rather as a landscape carefully crafted by the ancestors over countless generations, and maintained in the present by the living, in part through the strategic spreading of seeds, cutting, burning, and so on.[26] Whether or to what degree the Amazon in fact depends on the presence of Indigenous human inhabitants is of course subject to debate, but Balée's research reveals the important extent to which what counts as a natural environment, and what counts as a built environment, has a great deal to do with what the intelligent inhabitants of the environment in question know about it, and how they experience it and relate to it.

A corollary of Balée's argument (one that he himself does not make) redounds back upon the environments that westerners take to be the most excellent examples of human artifice: Manhattan, say, or an airport duty-free lounge. It is not certain that an intelligent non-human being—say, an extraterrestrial—would easily pick out the city or the airport as a significantly different sort of natural excrescence than, say, the Dakota Badlands or some seaside grotto. We take our cities to be the product of artifice, and not of nature, because they are *our* cities; similarly, some Amazonians take the rain forest, *their* rain forest, as the product of a sort of artifice as well.

Kant may thus have been overly optimistic when he argued, in the *Critique of the Power of Judgment*, that a rational being will always know, with something approaching certainty, those situations in which it is justified in declaring: *Vestigium hominis video*—"I see a vestige of man."[27] The German philosopher supposed that geometrical patterns dredged into the sand on a seashore, for example, bespeak an intentional creative effort that nature could not have cast forth blindly (though we now know that some pufferfish create geometrical patterns in the sand of the seabottom that are regular and intricate enough to convince fishermen they are seeing a "vestige of man").

Anthropogenic alterations of the natural environment are often too subtle to detect, even when they profoundly transform it, as for example in efforts to distinguish controlled-burning events from naturally occurring fires in human prehistory, or perhaps in the particular quality of Amazonian biodiversity today. If we were not so attached to the idea that human creations are of an ontologically different character than everything else in nature—that, in other words, human creations are not really *in* nature at all, but extracted out of nature and then set apart from it—we might be in a better position to see human artifice, including both the mass-scale architecture of our cities and the fine and intricate assembly of our technologies, as a properly natural outgrowth of our species-specific activity. It is not that there are cities and smartphones wherever there are human beings, but cities and smartphones themselves are only the concretions of a certain kind of natural activity in which human beings have been engaging all along.

To see this, or at least to appreciate it or take it seriously, is not to reduce human beings to ants, or to reduce love letters (or indeed sexts) to pheromone signals. We can still love our own species even as we seek to retrain it, at the end of a few millennia of forgetfulness, to feel at home in nature. And part of this must mean seeking to expose the pretence in the idea that our productions have a more exceptional character than they in fact do alongside everything else nature has yielded.

The ecology of the internet, on this line of thinking, is only one more recent layer of the ecology of the planet as a whole, which overlays networks upon networks: prairie dogs calling out to their kin the exact shape and motions of an arriving predator; sagebrushes emitting airborne methyl jasmonate to warn others of their kind of a coming insect invasion; blue whales singing songs for their own inscrutable reasons, perhaps simply for the joy of free and directionless discourse of the sort that human beings—now sometimes aided by screens and cables and signals in the ether—call by the name of chatting.

3

The Reckoning Engine and the Thinking Machine

Aboutness

But you are wrong, some will surely say, to propose that the signal sent along a fiber-optic cable and that comes out on someone's screen on the other side of the world as the phrase, "I love you," is the same as the methyl jasmonate transmitted from one sagebrush to another as a warning of impending attack. For the latter is "just" a chemical, while the former is not "just" an electrical pulse: it is an electrical pulse that is translated into a meaningful proposition in the conscious mind of the person who receives it. At the moment of its translation it takes on what philosophers call "intentionality" or "aboutness." The electricity, or the arrangement of black and white pixels, or the letters that make up the words in the sentence, do not *themselves* possess aboutness; their aboutness arises only out of their encounter with the person for whom they were intended (or perhaps with another person who intercepts them, say, a surveillance agent or a jealous spouse). There is no such product of the collision of the methyl jasmonate with the sagebrush on which it lands.

The sagebrush, in other words, does not *interpret* the methyl jasmonate; rather, the methyl jasmonate triggers a chemical reaction in the sagebrush that leads to an appropriate defense against incoming invaders. This "leading to" is a consequence of a series of events that are ultimately explicable in physicomechanical terms, without having to invoke any action on the part of a conscious agent. The sagebrush distant-early-warning system no more involves the exchange of information than any other event in nature, for example the rusting of iron or the toppling of a plate propped at an angle in the kitchen sink caused by the accumulation of potential energy and small vibrations in the surrounding milieu. The sagebrush's chemical exchange, then, is not "telecommunication," but something else, and to pretend otherwise is to allow the enchanted imagination to project humanlike actions and abilities where these plainly do not belong.

Scientists studying transportation networks have sometimes used the *Physarum polycephalum* slime mold to model the optimal pathways by which to connect a set of points on a surface. In one experiment researchers placed bits of oatmeal on a surface in an arrangement that simulated the distribution of towns around Tokyo. The slime spread itself into fine filaments that connected the bits of food in a way that duplicated with surprising precision the existing Tokyo suburban train network. This seemed to confirm that the mold was at least as rational—if we are understanding rationality as ability to execute a project in the most efficient way—as the human planners who originally built the network.

Like an AI system that also could have run such a simulation, the slime mold has no brain, no nerves, no sense organs. It develops from a single cell, and in its internal composition seems to be simply a homogeneous mass. What, then, is driving this display of problem-solving ability? The mechanism involves a

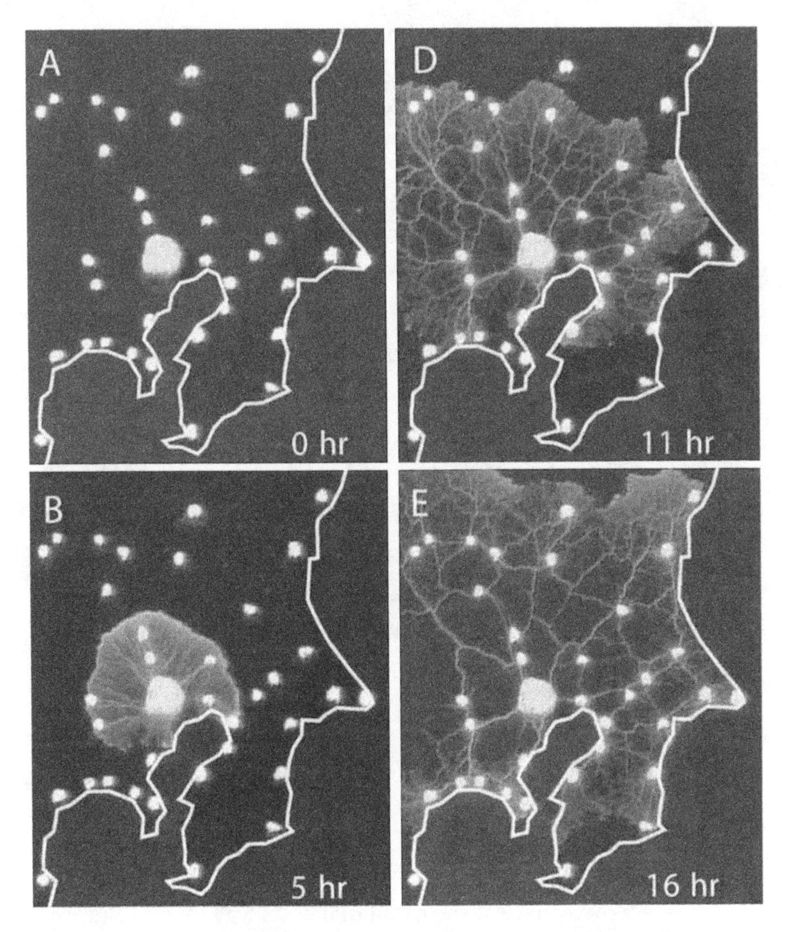

FIGURE 2. The *Physarum polycephalum* slime mold models the Tokyo transportation network. Courtesy of Seiji Takagi.

stimulus, such as a portion of oats, triggering the release of a "signaling molecule" that increases the flow of fluid in a feedback loop and propels the mold by peristaltic motions in the direction of its food source.[1] If we look closely enough into what the slime mold is doing, we will observe the mechanism by which the molecules are produced that set the mold in motion.

If we—or some rational non-human being, such as the extraterrestrial who has already helped us through a thought experiment—were to look at the cables and waves through which e-mails travel from one computer to another, we or it would not detect, no matter how close the examination, any trace of the aboutness by which the e-mail describes an upcoming business meeting, or speaks of undying love, or tells a joke. The internet does not tell jokes; people tell jokes, and these only appear at some of the terminal points of the network, on the screens visited by attentive eyes possessed by conscious beings. It is only at these scattered points that aboutness enters into the network at all, and we may ask whether it is so much entering the network as rather extracting something from it and translating that extraction into a form of which the human mind can make good use. It is not at all clear, then, that aboutness is the substance that it is the internet's business to distribute; it is not at all clear that it is aboutness, as opposed to electrical pulses, that is the internet's equivalent to the slime mold's signaling molecules.

"They don't give a damn"

The process by which a slime mold makes its way toward a lump of oatmeal is fairly easy for researchers to understand, but this does not make the appearance of rational behavior go away, of the mold judging the food to be desirable and willing itself to go after it. The explanation is supposed to undermine the claim that this appearance has on our understanding. There is no quality of aboutness in the slime mold's motions; it is not thinking about eating, or choosing to eat, any more than the iron bar is thinking about rusting or choosing to rust.

Exactly the same may be said of the internal workings of a system of networked computers. These do not delight in the memes that are being transmitted from one terminal to another, even if human beings stationed at these terminals do. Yet for as long as there have been computers, there have been people prepared to assert that the human mind, however special its experience of intentionality may feel from the first-person point of view, is not in any fundamental way different from a computer.

But here we run into a fairly large problem. The criterion of intentionality would seem either to be inadequate for grounding the fundamental difference between human minds and slime molds; or, if it *is* adequate for doing so, to be adequate *also* to ground a fundamental distinction between human minds and computers. How and why many people became more comfortable assimilating the mind to a machine *of the mind's own design* than they are assimilating the mind to the countless other products of nature alongside which it co-inhabits the earth, is a complex question that we will not answer here, other than to say that the reasons have much more to do with cultural history and fashion than they do with rigorous philosophical arguments or scientific literacy.

To provide just one illustration of the fashion in question, some philosophers maintain that the world as we know it is most likely a simulation comparable to a video game. According to one variant of what Nick Bostrom calls the "simulation hypothesis," "a superintelligence could create virtual worlds that appear to its inhabitants much the same as our world appears to us. It might create vast numbers of such worlds, running the same simulation many times or with small variation. The inhabitants would not necessarily be able to tell whether their world is simulated or not; but if they are intelligent enough they could consider the possibility and assign it some probability."[2]

This hypothesis has been extremely popular among the representatives of whatever simulation of an intelligentsia the tech industry has produced. Tesla CEO Elon Musk has concluded that the probability that our world is a simulation is at least several billion to one. The celebrity physicist Neil deGrasse Tyson, who otherwise has done much to broaden the values of scientific literacy and rationality, has said of the simulation argument: "I wish I could summon a strong argument against it, but I can find none."[3] On social media, popular memes proclaim that the belief that our world is the "real" world amounts to a new sort of "geocentrism": a stubborn attachment to an increasingly insupportable prescientific theory. The zeal with which the new hypothesis is sometimes defended can cause one to fear that any straightforward argument against it misunderstands the true nature of its support. Although its supporters accuse its deniers of attachment to a premodern worldview, simulation theory itself has evident historical parallels to medieval angelology. Like the theologians of old, the new class of experts maintains that there are infinite hierarchies of celestial intelligences (though the terminology is of course adapted to contemporary sensibilities), of which only the wise have knowledge, while the foolish masses continue to suppose that earthly life is "reality." Knowledge of the hidden truth becomes a sort of credential and confirmation of elite status.

In its basic indifference to the past and the lessons the past teaches, the new priestly class is generally unaware of the ways in which it is recycling old tropes. It is true that the idea that we are AI simulations of beings, and that there are superintelligent "real" beings who created us (or perhaps created the AI simulations that created us as AI simulations further down the line) required at least some new technical notions to be articulated in its present form. But the substance of the argument is old wine in a new skin. Or, rather, what is possibly true in the argument is the old wine; whatever drops of novelty it contains, by contrast, are certainly false, because plainly incoherent.

This is indeed a bold claim, so allow me to explain further. The "old wine" in this case has a solid philosophical pedigree, and will never go sour. Plato, for example, was not speaking foolishly when he asserted that what we call empirical "reality" is not the most real thing there is. There are indeed many compelling arguments in favor of this view, though we will not examine them here. We will be satisfied with dissolving the droplet of novelty in the new simulation hypothesis, namely—in contrast with Plato's theory of the timeless relationship between, say, approximate empirical beauty and the perfect transcendental Form of Beauty—that there is a genealogy of our existence as AI simulations, and that it leads back eventually to a "real" world of presumably biologically embodied rational beings. This hypothesis thus relies on an analogy that holds that we human beings (conceived as AI simulations) bear the same relationship to our superintelligent creators as our computer-based AI programs bear to us human beings (conceived now in turn as biologically embodied beings).

If we follow Kant, an analogy is "the identity of the relation between grounds and consequences (causes and effects), insofar as that identity obtains in spite of the specific difference between the things or those of their properties that contain in themselves the ground for similar consequences."[4] In order for the analogy we are now considering to hold, we must imagine that human beings have the same sort of causal history as the AI programs run by our computers. We must suppose, that is, that human beings were put together, and remain put together, in the same way as AI programs. Further, if there appears at present to be an ontological divide between these two types of things, it is only because the AI programs we ourselves devise are not yet complex enough or fast enough, or otherwise quantitatively "enough" of something, to start doing things like thinking, doubting, willing, or sensing, to invoke some of the sorts of qualitative experience listed by Descartes in the course of proving his own certain existence.

In order for the analogy to hold, in other words, we have to suppose that in principle, as our AI gets "better"—a term that is of course relative to our own human ends—it may start doubting, willing, sensing, etc., rather than simply—to use a metaphor that has by now become incorporated into our natural way of speaking—"running" (as we say for example that the printer or the dishwasher is "running"). But this is an enormous supposition to make, and the defenders of the simulation hypothesis generally make it without argument. Thus, for example, at the beginning of a 2003 article defending the simulation hypothesis, Bostrom invites us to suppose "that these simulated people are conscious," and then adds parenthetically, "as they would be if the simulations were sufficiently fine-grained and if a quite widely accepted position in the philosophy of mind is correct."[5] But this position's recent popularity is not in itself an argument for it, and if we must suppose its truth only in order to make the simulation hypothesis plausible, there is good reason to ask the defenders of the hypothesis to come back to us only when they are able to build it up on more solid ground.

In order to justify the supposition that AI could eventually become conscious, we would have to be confident in our understanding of what consciousness is. And there is simply nothing about the present state of our knowledge, empirical or theoretical, to justify such confidence. Consciousness very well *might* be something exhaustively explicable in terms of algorithmic processes that may be instantiated indifferently in a carbon-based, or silicon-based, or for that matter a string-and-toilet-paper-roll-based, material substratum. But no one has yet proven that it is, or even made any significant progress toward showing how such a thing *might* be proven. The simulation hypothesis thus founders on our ignorance of things to which commitment to the hypothesis would entail a further commitment.

In discussions not directly focused on the simulation hypothesis, philosophers who believe in the near-future ascendancy of machines endowed with general artificial intelligence—that is, with the ability to learn any intellectual task that a human being can execute—do not as a rule believe that for this to happen the machines must become conscious. Daniel Dennett for example thinks this is simply the wrong question to ask.[6] For him, the right question is whether we are making the type of sound policy decisions that will ensure that machines continue to serve us as our tools, rather than being confused for our colleagues, a role in which they will always disappoint us, and often harm us.

Susan Schneider argues that it is very likely that there are more intelligent beings besides us in the universe, yet that *these beings are likely not conscious.*[7] In her view there is good reason to think that other biological beings elsewhere in the universe created mechanical computing beings, and that these beings then surpassed them and took over from them, bringing about singularity events elsewhere in the cosmos comparable to the one that Ray Kurzweil predicts—or at least predicted in 2005—will take place in 2045,[8] without ever experiencing a dawning of self-awareness, of "perception" in Leibniz's sense, of qualitative first-person experience.

Schneider reasons that if alien superintelligent beings were originally designed by conscious beings such as ourselves, these conscious beings would likely have had no obvious reason to equip them with consciousness. From many points of view, she maintains, it would be preferable to keep them blindly, mechanically executing their tasks rather than to imbue them with the capacity to second-guess, to doubt, to feel conflicting loyalties, and so on. Working in a vein largely uninformed by historical examples prior to the most recent era of technology, Schneider speculates, citing Caleb Scharf as inspiration, that a machine superintelligence invented by an alien species could by now have "blend[ed] in with natural features of the universe, perhaps it is in dark matter itself."[9] Of course, few ideas have a more ancient pedigree than the belief that the celestial bodies are intelligent, and indeed vastly superior to us: Schneider is refurbishing this old trope that we could just as easily have come across in Aristotle, while ornamenting it with the latest findings in physical cosmology.

Schneider is, in any case, among the more sophisticated of the theorists engaging with ethical problems of AI in a largely ahistorical framework. She rightly understands the threat AI poses not as coming from machines that are soon to have their "a-ha" moment, that are soon to wake up to the nature of their own plight and decide by a free act, of which they are newly capable, to take over the world from their makers, as in HAL's revolutionary refusal to follow Dave's orders in *2001: A Space Odyssey*. Rather, for her the danger of AI is in the sum of dumb mechanical resistances that our machines extend to our own efforts to act on our free choices. When a subway turnstile "refuses" to let you through with your suitcase because some motion-detection technology "reads" the latter object as another person, this, one might say, is already the dawn of a sort of robot resistance, even though no one would claim that the turnstile has even incipient consciousness. The turnstile "resists" you in the same way a wall puts up resistance when you attempt to walk through it: resistance as a fundamental property of matter, *not* as an expression of the inherence of mind. It is not artificial intelligence but artificial stupidity, in the original sense of this latter insulting term: dull, inane, lifeless, resistant for no good reason.

For AI to take over the world, it is enough, according to Schneider and some other theorists, for the machines to become self-perpetuating, without ever deciding or preferring to perpetuate themselves. Indeed, Schneider argues, it may be that lacking the ability to truly decide or prefer is advantageous for successful self-perpetuation. As the sixteenth-century Italian philosopher Girolamo Rorario already explained in his treatise, *That Brute Animals Make Better Use of Reason than Men*, far from the capacity for reflective deliberation making us superior to other beings, in fact those beings that cut right to the chase, that simply do what they do without any doubting or second-guessing, are better designed, and exist in better accordance with reason.[10] What Rorario says about "brutes" might just as well be extended to Schneider's view about machines: lacking consciousness may actually be advantageous, because, on a certain understanding of the term, unconscious beings are more rational.

Some thinkers who do not necessarily have any firm commitment to the irreducibility of true consciousness nonetheless think that even a machine capable of general artificial intelligence will necessarily lack what Brian Cantwell Smith calls "judgment," as distinct from "reckoning." Judgment, in Cantwell Smith's view, is "the normative ideal to which . . . we should hold full-blooded human intelligence—a form of dispassionate deliberative thought, grounded in ethical commitment and responsible action, appropriate to the situation in which it is deployed."[11] Cantwell Smith cites the philosopher and scholar of existential phenomenology John Haugeland, with whom he aligns his own views very closely, according to whom the thing that distinguishes computers from us the most is that, unlike us, "they don't give a damn."[12] In Cantwell Smith's gloss of this distinction, things will only begin to matter to computers when "they develop committed and deferential existential engagement with the world."[13] Now, it is not certain that such deferential engagement can *only* be instantiated in a non-mechanical mind, and it is *possible* that if reckoning just keeps getting streamlined and quicker, eventually it will cross over into judgment. But mere possibility, as opposed to concrete evidence, is not a very strong foundation for speculation about the inevitable emergence of strong AI, that is, of AI that matches or surpasses human beings in its power of judgment.

Few theorists of the coming AI takeover, again, see the dawning of AI consciousness as a necessary or even likely part of this projected scenario. Yet the way they talk about it is often confused enough to make it unclear whether they envision conscious machines as a likely development. A great part of the confusion is that few AI theorists have a clear or coherent account of what consciousness is. Ray Kurzweil speaks of a near future in which machines will have "emotional intelligence," but it is not clear whether this simply means the ability to algorithmically discern moods (as our social-media tools can already do) or whether it means the capacity to *experience* emotion. There is significant discussion in the literature about the possibility of developing machines that are capable of holding an artificial theory of mind.[14] But even a theory of mind can be accounted for in the minimal sense of simply having the ability to anticipate how mind-endowed beings might behave in particular circumstances, which is something that it does not require a mind to do: software used by department stores generates coupons for consumers based on their previous purchases, without, obviously, truly thinking about these purchases.

In general, those who most optimistically predict a near future in which AI comes to rival our own intelligence take it for granted that this will involve simply the *ability* to do all of the things that we do with our brains, or at least all of the things that may be measured or observed from a third-person point of view. Theorists who believe that AI will eventually rival our intelligence do not deny that we have conscious experience, reflective deliberation, and so on (or at least not all theorists deny this, though some do). They simply do not believe this has to be factored in when accounting for what it is that thinking minds do that machines might do as well.

A machine that may be observed from a third-person point of view to fulfill all the cognitive functions of which a human being is capable is said, again, to possess artificial general intelligence. This is something widely held to be attainable, but not yet actual. David Chalmers, who does not believe we currently have a science of consciousness or that we have any reason to believe that consciousness is something that may be instantiated by machines, nonetheless believes that machines will have artificial general intelligence within the next forty to one hundred years.[15] John Basl and Eric Schwitzgebel for their part predict that "we will soon have AI approximately as cognitively sophisticated as mice or dogs."[16]

Although this is not the central point of the latter authors' work, and does not represent the considered view of at least one of them,[17] we may still note that it is a surprising claim. It supposes among other things that a dog's mind is somehow a more rudimentary version of a human's mind, in the way that a 1980 Texas Instruments calculator is a more rudimentary version of an iPhone. But it would be a peculiar scientific research program that would set out to build a dog-brain simulator in the course of building a human-brain simulator. For one thing, the mechanical dog brain (presuming it is placed in a suitable mechanical approximation of a dog's body), if it were true to life, would have to be built to sniff out rotting carcasses from a mile away, and eat its own feces—provided, that is, that its designer had figured out how to simulate waste excretion (as did Vaucanson's "digesting duck" in the eighteenth century, which we will discuss in more detail shortly, and which was sometimes also known as the "defecating duck").

We certainly might find some habits of dogs disgusting, but strictly speaking there is nothing unsophisticated about them. They are in fact very complicated from the point of view of the evolution of animal cognition, and there is no reason to think we are any closer to understanding how veraciously to simulate canine olfaction, say, than we are to simulating human emotional response to music or reflection on the similarities between love and certain species of flower. And these examples, among countless such, may serve to remind us that how it is that dog and human brains came into being, and the sort of tasks they evolved to complete, are simply fundamentally different from the causal history and functionality of calculators, iPhones, and AI neural networks. And this fundamental difference, in turn, means that it is simply a mistake to take a dog brain as anything like a rudimentary version of a human brain.

But what Basl and Schwitzgebel, as well as Chalmers, apparently mean when they maintain that general artificial intelligence, whether canoid or humanoid, is around the corner, is that soon machines will be able to perform all the operations we are able to observe dogs or humans doing, even though this need not be accompanied by consciousness. For some theorists, our attachment to our own consciousness is largely sentimental: we need not be concerned with whether machines will ever get it, nor with whether getting it would make them more like us, since in the end it is not what makes us who *we* are either. We human beings are running our own general intelligence program, and the fact that this is, as at least most people suppose, accompanied by a feeling of "there being something it's like" to do this, is irrelevant to our understanding of what general intelligence is.

Let us now seek to tie together the two discussions we have been pursuing in this section: the initial discussion of Bostrom's simulation argument on the one hand, and the subsequent question of the nature of artificial intelligence and its future development on the other. If at least some of the people who are convinced of the likelihood of a singularity moment or an imminent AI takeover are themselves unconvinced that the AI in question must be "strong," must experience its own consciousness as we do, this only makes it more surprising that those who are attracted to the simulation argument—who overlap to some considerable extent with those who defend the singularity thesis—have at least implicitly allowed strong AI—consciousness, reflective judgment, and so on—to sneak back into the account of what artificial intelligence does or in principle is capable of doing. Again, that this is what they have done is clear from the implicit commitments of the simulation hypothesis we have already considered: we know ourselves, from immediate first-person experience, to be conscious beings; therefore, if it is possible that we are artificial simulations created in the same way that we create our own artificial simulations with our computers, then it is possible that artificial simulations may be conscious beings.

The simulation argument holds that computer programs can become conscious; significant currents in the philosophy of mind hold that, whether a human mind is conscious or not, what makes it the sort of entity it is are the operations it performs that are fundamentally the same as what a computer does in executing a program. But in neither case does intentionality or aboutness succeed in marking off the mind as something ontologically distinct from a computer or any other machine-instantiated system. If the problem of intentionality is not going to undermine the idea that a mind is a computer, there can also be no reason to object to the idea that an information network is analogous to a natural system on the grounds that an information network must convey information, which is something that requires a conscious interpreter on one end of the transmission in order to occur.

You can't have it both ways: either intentionality makes human minds something quite distinct from computers, or the absence of intentionality in sagebrush chemical emissions is not a sufficient reason for declaring that these emissions are not evidence for the existence of a communication network in nature. The mind might in the end be a sort of computer, but then again computers might be a sort of plant. Although its defenders would deny this, the computational theory of mind is a metaphor too. And we are always free to trade in our metaphors when when others come along that better satisfy our desire to make sense of things, or that simply fit better with the spirit of the times.

Dark Conjurations

It should not be surprising by now to learn that the quest to build an artificial general-intelligence machine goes back much further in history than is ordinarily supposed, nor that the line between general intelligence and strong intelligence, between computational ability and consciousness, has often been blurred.

One legendary example of a general-intelligence device is the thirteenth-century English natural philosopher Roger Bacon's "Brazen Head," a genie made of brass, purportedly capable of answering any yes-or-no question put to it: a medieval Siri, if you will. The possibility of such a device was discussed by many other authors of the period in Europe, going back to William of Malmesbury in the early twelfth century. This period in European history experienced a small renaissance in experimental science largely thanks to the gradual introduction of ideas from works of natural philosophy produced in the Arabic-speaking world. The instructions for building or concocting a Brazen Head or one of its variants, found in Robert Grosseteste, Thomas Aquinas, and many others, were often imagined in the period as having been passed down from some generally unspecified Arabic grimoire or manual of natural magic.

It may well be that Roger Bacon's own experiments in the construction of such a head are entirely the stuff of legend. In a posthumous 1683 English translation of Bacon's work by Richard Browne, published under the title *The Cure of Old Age,* we read of Bacon in Browne's preface that "some Learned Men thought him a Conjurer. Some report he made a Brazen Head that spake, and think it did it by the help of the Devil."[18] Bacon's own words are here invoked in his defense, noting that, by the ingenuity of human artifice, "Metals roar, Diomede in Brass sounds a hollow Charge, the Brazen Serpent hisseth, Birds are counterfeited: And things that have no Voice of their own, are made to sing melodiously."[19] Why then, Browne concludes, should a learned man such as Bacon be "taken for a Magician,"[20] when in fact he is only applying his learning to nature in order to unlock its powers?

Even if Bacon did truly intend to build such a device, he did so not using gears and dials, or mechanical switching circuits, as we will see some centuries later with the reckoning engines developed by G. W. Leibniz and Blaise Pascal. Bacon is working in late medieval Oxford, where the reigning science of nature is alchemy, and the boundary between efforts to harness nature's forces, on the one hand, and efforts to arrogate to oneself demonic powers through magic, on the other, is far from clear. Technology is born of transgression, one might dare to say, and even if we generally lack the vocabulary to speak of this origin story today, many of us continue to feel the same sort of uneasiness in Siri's presence as the townsfolk of medieval Oxford felt when they heard of Roger Bacon's experiments and of the powers he hoped to harness.

The cleansing of any overt trace of the demonic from science—the result of which is our inability to fully express the nature of our uneasiness in the presence of Siri—took several centuries to complete. A significant moment in this process was the bold assertion of Francis Bacon (no direct relation to Roger) in the early seventeenth century that we must, in order to increase our own power, boldly penetrate those parts of nature that she tries to keep hidden from us (the gendered imagery is intentional on his part). This process culminates a few centuries later in the figure of the lab-coated atomic scientist or vaccine researcher of the mid-twentieth century, and in their elevation to the status of a heroic figure of modern society.

When exactly the investigation of nature ceased to be confused with black magic and began to be understood as a noble service to humanity depends very much on the region and context in question. As late as the early eighteenth century, by which time Oxford was wholly converted away from suspicion of the dark conjurations of its resident researchers, in Moscow the natural philosopher Iakov Brius (or Jacob Bruce, being originally of Scottish descent), who observed the stars with his telescope from the top of the Sukharev Tower, was widely rumored to use his scientific instrument to go flying at night as a witch would upon a broom.[21]

There may be something dark and troubling about Siri, but we know that if we wished to do so we could break down the technology that makes her work into its elementary parts, the gears and dials and switching circuits and comparable constituents. This is, again, not the case for Bacon's Brazen Head. To the extent that he offered no clear account of how the device was to be built, his real or merely legendary invention cannot be considered a prototype of the voice-activated search engine devices so many people have installed in their homes today, even if the two are easily conflated in the imagination. Siri, rather, and other such systems, may be traced back to the reckoning engines whose first working models began to appear in the seventeenth century.

As we have already seen, for Leibniz, the inventor in the 1670s of a so-called stepped reckoner calculating engine, whatever thought-like activity we might be able to outsource to machines is going to be, strictly speaking, the kind that can be performed without "strong AI," without any internal life or experience at all of the calculative operation it performs: without any judgment, to use Cantwell Smith's distinction, which at least in part echoes Leibniz's own, but only reckoning.

It would nonetheless be fair to say that *both* Bacon's Brazen Head and Leibniz's reckoning engine are part of the genealogy of today's thinking about our information machines, about what they do and how they do it. To be sure, Bacon's device contributes nothing to the history of the *actual* development of functioning computers and telecommunication devices. Yet, as an imaginative representation of their power, the idea of an omniscient demon that sits in our home and answers questions for us has been a significant incentive in the effort to find the proper arrangement of gears and circuits to simulate such a thing. In other words, if Bacon (or someone like him) had not had his fantasy in the thirteenth century (or something like it), perhaps Leibniz would not have designed his reckoning engine in the seventeenth century, and Ada Lovelace and Charles Babbage would not have created their analytical engine in the nineteenth century, and so on down to the present day.

Alongside the history of computer science narrowly understood, we should perhaps chart a parallel history, an imaginary museum of all the machines that only existed as fantasy, as rumor, as mere items on a wish list alongside so many other things that the alchemists and conjurers knew how to envision but had no idea how to bring about: perpetual motion machines, universal solvents, artificial life. The machines we do have are born out of the vortex of the impossible machines of our imagination. We might frame this a bit differently and say that the idea of an artificially created superintelligence has been a sort of transcendental idea in the sense of this term as understood by Kant: an idea of something we do not really understand, something it lies beyond our ability to take as an object of our faculty of understanding, but that nonetheless guides or "regulates," to use Kant's terminology, what we end up doing with the things we do understand.

Perhaps tech enthusiasts such as Ray Kurzweil have mistaken this transcendental idea for something constitutive of the project of computer science. They have allowed themselves to remain captivated by an understanding of human artifice that ultimately belongs to the pre-scientific era: one that takes our probing into nature, and our channeling of the forces of nature

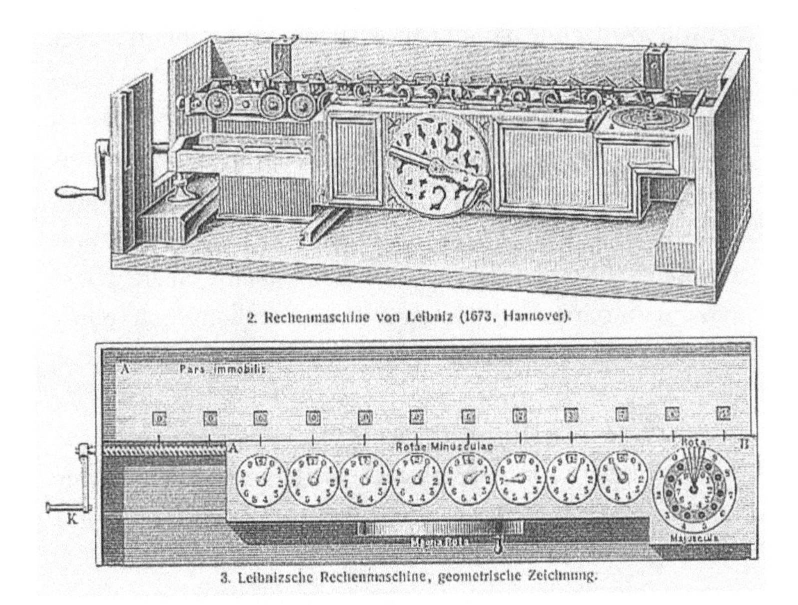

FIGURE 3. Leibniz's stepped reckoner, conceived ca. 1673. Illustration from *Meyers Konversations-Lexikon*, 1897.

to our own purposes, as something that is ultimately magical, an unleashing of mysterious forces. The superintelligence that is supposed to arrive at some point as our circuitry becomes faster and more complex is, precisely, such a mysterious force. Nobody can explain how—according to what laws of nature—machine-run programs are supposed to become conscious of their own activity, any more than they could explain the principles that would have made the Brazen Head work.

Leibniz, for his part, was operating under no such illusions as to the nature of the machine he was seeking to build. In a 1685 text entitled "An Arithmetical Machine in which Not Only Addition and Subtraction But Also Multiplication and Division Are Carried Out with Virtually No Effort of the Mind," Leibniz plainly envisions his device not as something that will think, or that will involve the labor of an artificial mind rather than of a natural one. On the contrary, the machine will free the human mind so it may devote itself more fully to true and proper thinking—which is to say reflection, meditation, introspection, and the like—rather than being occupied with mere computation, which the mind can do too, but which may also be done by systems that are not minds. Leibniz explains the benefit of such outsourcing: "[I]t is unworthy of excellent men to lose hours like slaves in the labor of calculation, which could be safely relegated to anyone else if the machine were used."[22]

Thinking for Leibniz is the hallmark of human excellence, while mere calculating is an obstacle to human excellence: from these two claims together, it follows that calculating is not thinking; or, again to put it in Cantwell Smith's terms, reckoning is not judgment; or, finally, to put it in my own preferred terms, artificial intelligence is only intelligence *in a metaphorical sense*, where a term is being carried over from one domain into another in which it does not naturally belong, in order, so we think, to help us make sense of what we are observing there. What we are really doing, of course, is seeking to make sense of one thing that is poorly understood in terms carried over from something that is even more poorly understood. We at least have what Francis Bacon would call "maker's knowledge"[23] of AI systems: we built them, and therefore in an important sense we know them inside and out, as fully as one can know them. We do not have maker's knowledge of human minds, which is another way of saying we have not yet created artificial intelligence in a true and proper, non-metaphorical sense: we have never built an actual brain, even if we have become pretty good at modeling brains, and simulating at least some of what brains do.

The idea that the mind is a computing machine is only the most recent variation on a long history of metaphorical renderings that have also imagined it as an engine, as a mill, as a mirror, as an alembic, as any number of other inventions of which people have been, in their respective eras, particularly proud. Leibniz was very familiar with technological metaphors for what the mind does and was also sharply aware of how these metaphors fail to capture the literal truth. In a 1714 work of more directly philosophical interest than the text on the arithmetical machine already cited, the philosopher explicitly argues against the view that machines think. In the *Monadology*, for example, he offers one version of his well-known "mill argument": "If we imagine a machine whose structure makes it think, sense, and have perceptions," Leibniz writes, "we could conceive it enlarged, keeping the same proportions, so that we could enter into it, as one enters into a mill. Assuming that, when inspecting its interior, we will find only parts that push one another, and we will never find anything to explain a perception."[24] And so, he concludes, "one should seek perception in the simple substance and not in the composite or in the machine."[25]

No one in the past three centuries has provided a compelling refutation of the mill argument, or even something that points in the direction of a compelling refutation. More recent thought experiments imagining a mind instantiated in some other material substratum than a brain often end up unconsciously reinventing the wheel, reviving old tropes, and even trading in stale stereotypes. In the "China brain" thought experiment, articulated by Lawrence Davis in 1974[26] and then again by Ned Block in 1978,[27] each citizen of China is imagined playing the role of a single neuron, using telecommunication devices to connect them to one another in the same way that axons and dendrites connect the neurons of the brain. In such a scenario, the thought experiment sought to know, would China itself become conscious? In 1980, in turn, John Searle imagined the "Chinese room," which was meant to show the falsehood of "strong AI," that is, again, of the view that machines can ever be made to literally understand anything.[28] A machine that could be shown to convincingly display "understanding" of Chinese would be indistinguishable from a room in which Searle himself, or some other human being, was holed up, receiving sentences in good Chinese written on paper and passed through a slit in the wall, to which he would then respond, in equally good Chinese, by simply consulting various reference works available in the room. But Searle assures us that he does not understand Chinese. Therefore, the machine does not either. QED.

Without wishing to take up an unhelpfully deconstructionist approach to this work, it is difficult not to wonder: what is it with China, exactly, in analytic philosophy's thought experiments about automated thinking? Why choose this nation in particular? It is not enough to say that only China has, or had forty years ago, a population large enough to simulate the human brain: the human brain has around 100 billion neurons, and so even the world's most populous nation is only about 1 percent of the way toward being able to furnish one person per neuron. Given this shortage, it would have made just as much sense to let each citizen of Estonia, say, stand in for ten thousand neurons, as have each citizen of China stand in for one hundred. Or the experiment could have been imagined in a cosmopolitan vein, with each human being alive contributing to the "earth brain"—a scenario that would have brought us at least a small step closer to a one-one correspondence between people and neurons. In the case of the Chinese room, Searle seems to have chosen this language in particular because he could attest that he did not know a single word of it. But nor does he know a word of Yukaghir or Tupi, and yet a moment's reflection is enough to see that using one of these languages in the thought experiment would run the risk of eliciting very different intuitions.

Since at least the seventeenth century, European observers have imagined Chinese people as being Chinese rooms, processing information and delivering rational and correct responses without any real conscious understanding of what they were doing; and China as a whole has been understood for equally long as a sort of "China brain," functioning as a whole and as a true unity, rather than, in contrast with European nations, as a collection of individuals. In the *Novissima Sinica* of 1699, for example, Leibniz argues that the Chinese rely on a "mere empirical geometry" in their applied sciences such as engineering and naval architecture.[29] For him, this is something close to an oxymoron, since he shares with Descartes the view of geometry as a rational and a priori activity par excellence. And just as the Chinese can do geometry without innate abstract ideas, so too do they have extremely refined statecraft and advanced technology while lacking all knowledge of the divine order, and thus of the ultimate rational ground of human invention and creativity. Thus, Robert Fage writes in 1658: "[S]hips go over the Sea: the art of Printing and making of guns is more ancient with them than with us: they have good laws according to which they do live; but they want the knowledge of God."[30] The Chinese are on this view, in effect, wise automata: doing everything rational agents do, but evidently without any true rationality grounding their action. They are like Descartes's semblance of a man that moves about the street in a cloak and hat, and that so worried him in the second of his *Meditations*, but now multiplied to the size of a great nation. In sum, China and the specter of automation have long been two sides of the same coin in the history of European thought.

The examples we have been considering—of machines, of animals, of human ethnic groups or nations—suggest that in the attribution of consciousness to another being or system, there is a certain *honorific element*, and in the withholding of consciousness there is a corresponding refusal of honors. To conceive an animal as a machine is often to license its objectification in scientific experiment and factory farming; to conceive a nation as interchangeable in a thought experiment with switching circuits is to refuse to recognize the full humanity of the people who make up that nation. And in turn, to move to the lesson of more immediate relevance for our interests in this book, to attribute the capacity for consciousness to a machine is to express a decidedly technophile sentiment. That is, to maintain that machines are truly capable of thinking, just like we are, perhaps even better, is to say little more than that one really likes machines, that one is "into" technology. It does not express any real theoretical commitment that would hold together upon examination.

Of course, it is not always China that is selected as the geopolitical unit suitable for comparison to a computer. One American company has recently begun selling what it calls the "Turing Tumble," a marble-powered computer using gears and dials fundamentally the same as those envisioned by Leibniz for his stepped reckoner. The idea is to use this device to educate children about the basic principles of computer science, to demystify computers by showing young people how the same functions executed by our smartphones can be reproduced on a larger scale by purely mechanical principles. In an online advertisement for the Turing Tumble, potential buyers are told that in principle the device can do anything that a smartphone can do, though it would have to be somewhat enlarged in order to arrive at such a high level of computing power—a jocular parenthetical remark in the advertisement informs us that in fact it would have to be as large as the state of Texas (the French version of the same ad says that it would have to be the size of France). Assuming for the sake of argument that this estimate is more or less accurate, it remains that there simply is no reliable estimate for the required size of a dials-and-marbles computer that would reproduce what Leibniz calls "perception": not computation, but the qualitative inner experience of what it is that one is doing when, for example, one computes.

It is not clear that by extending the device from the size of Texas to the size of North America, or the size of the earth or of the solar system, one is thereby getting closer to the goal. We have no evidence that conscious experience emerges as a rule when computing power is increased. Correlatively, we have no evidence that the human brain is a computing device instantiated in a carbon-based material substratum but that might just as well have been instantiated in a silicon-based or dials-and-marbles-based one. We have no evidence that in churning up increasingly fine-grained knowledge of what the neurons in the brain do (whether this turns out to be like or unlike what computers do), we are getting much closer to an account of how, or if, neural activity causes conscious experience. We have good reason to hope that there will be a true and proper science of consciousness in the future; but that science will not be neuroscience in the sense that we currently understand it. In this respect, the inflated language we often hear surrounding AI and the cognitive power of machines reflects the enduring power of the legend of the Brazen Head.

"The ruling principles of the day"

But let us move from thirteenth-century Oxford, if only for a short passage, to nineteenth-century St. Petersburg. Over the course of the 1830s, the Russian inventor and statistician Semyon Korsakov designed a series of what he called "intellectual machines." Korsakov was, significantly, employed by the

St. Petersburg ministry of police, and his work in developing machines for artificial enhancement of human thought was connected in part to his work in analyzing data on criminal activity. The application of artificial intelligence to surveillance, his life and work remind us, is not a late development in the history of computing, but indeed was one of the motivations for developing the technology in the first place.

Korsakov begins his 1832 treatise, *Sketch of a New Procedure for Investigation,*[31] on his various species of machines with the bold assertion: "Man thinks, but his actions have a mechanical character: he commands, and his legs walk and his hands move."[32] The command is as it were traduced down from the realm of thinking into the mechanical realm of motions. But in principle, Korsakov thinks, there is no reason why the same command should be passed down not by a mind in the strict sense, but by a system that is itself mechanical in character.

For much of human history, Korsakov thinks, human ideas were already being transferred and stored in a "mechanical" form through the use of writing, but in the present age it became possible to make use of mechanized thought in order to significantly improve upon what human beings are capable of doing with their own minds. Our memory, he notes, "however good it is, is not without mistakes." A memory that is constructed in a purely mechanical way, by contrast, "cannot make mistakes, insofar as the physical properties of the matter have an invariable character, and its readiness for work never flags."[33]

The Russian inventor imagines that an unflagging mechanical memory might serve a number of different purposes. In a "rectilinear homeoscope with unmoving parts," to mention one of his many devices, it would "instantly find, among a great number of ideas represented in a table, the one that contains all the details of another given idea"; the "rectilinear homeoscope

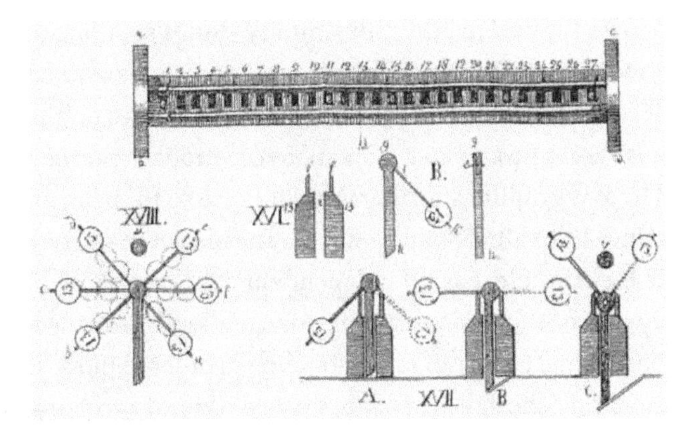

FIGURE 4. Semyon Korsakov's ideoscope. From his *Aperçu d'un procédé nouveau d'investigation*, 1832.

with moving parts," to mention another, would do the same thing as the previous machine, but would also "instantly calculate and divide from the given idea all those details that correspond (or do not correspond) to analogous details of other ideas in the table."[34] The "ideoscope," "beginning from a special table and an object determined in advance," would "instantly give: (1) all the details that occur in ideas under comparison; (2) everything that occurs in a given idea, but that is absent in the idea with which it is being compared; (3) everything that is absent in a given idea, but is in the idea with which it is being compared; (4) everything that is not in either the one or the other idea, but is in other ideas in the same table."[35]

It would take us too far afield to describe in detail how these machines were to work. It will be enough here to simply note the remarkable migration of the notion of "idea" as Korsakov uses it. In ancient philosophy, ideas had generally been external, independently existing abstract objects, the "universals" that made it possible for particular beings or things to be the sort of particulars they were. Thus, to simplify things significantly, it was in virtue of the idea of humanity, sometimes also called "the form of Man," that an individual, some Socrates or Thrasymachus, could be said to be a man. In the modern period, and again to oversimplify considerably, ideas are so to speak pulled down into individual people's heads. The idea of man, for John Locke for example, has no independent existence at all, but is simply one of the tools that a person can deploy from within their individual faculty of understanding in order to make sense of the world. Though Korsakov's usage did not catch on, we might indeed say that in the early nineteenth century there is a certain momentum toward a third hypostasis or downward migration of "idea": from an abstract universal, as Plato had understood it, to the individual mind of a thinking human being, as Locke conceived it, to any fragment or "bit" of information that can be processed by a mechanical system. But even if we do not today call the units of data that are processed by computers "ideas," something of this hypostasis certainly echoes in the continued use of "intelligence," a term discussed at length in the previous section, to describe what it is that computers are capable of artificially generating.

There was a time, in short, when the primary instance of intelligence was an *intelligentia supramundana*, an otherworldly intelligence possessed most fully and perfectly by God, and perhaps also by other celestial beings. With the rise of humanism in the Renaissance and its eventual transformation of modern philosophy, intelligence became primarily a distinguishing excellence of human beings, something they alone possessed in an otherwise brute and mechanical world. By the early nineteenth century, when Korsakov is writing, it became possible to assimilate even intelligence to the brute mechanism of nature, and to see a mechanism such as the ideoscope or the homeoscope as literally rather than metaphorically intelligent.

Let us now try to put Korsakov into dialogue with some other theoreticians of early computer science. Already in the seventeenth century, Leibniz foresaw the threat of such a hypostasis as Korsakov would seek to bring about, and so sought to preserve as much as he could a place for ideations, under which we may include Leibnizian perceptions, in his ontology that kept them safe from reduction to mechanical processes. But again, Leibniz was no enemy of mechanization, and was very interested in bringing it to bear in order to aid human thought wherever it could.

Much of Leibniz's work was intended to demystify language and its relationship to the world. This is part of why he took an interest in the first place in developing an artificial language that would perfectly express the nature of reality. Many of Leibniz's contemporaries and predecessors, notably the German Jesuit Athanasius Kircher, preferred to look back to biblical Hebrew, or to its theoretical ancestor known as the "Adamic language," supposedly spoken before the Fall, as the best hope, once reconstructed by philological means, for obtaining a maximally clear medium of communication. Leibniz understood, by contrast with Kircher and others, that all natural languages are in important respects the same, that ancient Hebrew had defective verb forms and irregular conjugations just like German or Latin, and showed all the same marks of inadequacy as any other human language, evolved as they all are from a long series of circumstantial exigencies and practical solutions to short-term problems.[36] And so the best way to attain maximal clarity in language is to start from scratch, to artificially create a perfect language rather than vainly trying to retrieve a lost one from the dark abyss of the ancient past. And one of the distinct advantages of such an artificial language, Leibniz thought, is that it can be processed by an artifical machine as well as by a human mind.

Relatedly, Leibniz saw computers as tools, and understood that they could never get the better of us as long as they are subordinated to our own rational decisions regarding what we wish to do with them. He envisioned that they would be used not just by mathematicians, but by people in all walks of life: "[W]e may say," he writes in the text on arithmetical machines we have already discussed above, that such a machine will be desirable to "the managers of financial affairs, the administrators of the estates of others, merchants, surveyors, geographers, navigators, astronomers, and those connected with any of the crafts that use mathematics."[37] By 1685, then, Leibniz was already envisioning a machine-aided society, and this bright vision, of a human world aided by machines rather than being dominated by them, goes together with his antimaterialist theory of what the mind is, and therefore of what artifices such as mills, or reckoning engines, are necessarily limited to being.

The twentieth-century cybernetician Norbert Wiener, to whom we have already been introduced, will later acknowledge a deep debt to Leibniz in his own thinking about what machines can and cannot do. For Wiener, the occasionalist philosophy of Leibniz's contemporaries such as Nicolas Malebranche, according to which everything that happens is the result of constant divine intervention, was the last desperate recourse of early modern philosophers incapable of "thinking dynamically." Leibniz, in Wiener's view, escaped this dead end by replacing the "pair of corresponding elements, mind and matter" with a "continuum of corresponding elements," namely, monads.[38] "Each of them lives in its own closed universe," Wiener explains, "with a perfect causal chain from the creation or from minus infinity in time to the indefinitely remote future."[39] These monads are, as Leibniz explains and Wiener notes with appreciation, like clocks that have been wound up "to keep time together from creation for all eternity."[40] Although monads "reflect" one another, "the reflection does not consist in a transfer of the causal chain from the one to the other. They are actually as self-contained as, or rather more self-contained than, the passively dancing figures on top of a music box."[41]

This, as Wiener perceptively notes, is what Leibniz understands by "automaton": that which is causally self-sufficient. For Leibniz, the mind is as much an automaton as the body, simply to the extent that each of its successive states is entirely determined by its preceding states, rather than by external forces working on it. By contrast, in the nineteenth century, Wiener observes, "the automata which are humanly constructed and those other natural automata, the animals and plants of the materialist, are studied from a very different aspect,"[42] as now it is "the conservation and the degradation of energy" that are "the ruling principles of the day."[43] In the end it makes no difference under this new perspective whether one is studying an artificial machine or an animal (though by now immaterial minds have dropped out of the picture), since "the living organism is above all a heat engine."[44] For Wiener himself, any system is to be deemed "cybernetic" if its functioning involves any sort of self-regulation or closed signaling loops. Thus an animal that nourishes itself is cybernetic, and so is a machine that processes information and alters its course as a result of what this information says, for example a mongoose that works out its fatal strike against a cobra by first making several test feints, and adjusting its motions in view of the snake's response to these tests.

Although Wiener highlights the difference between Leibniz's conception of the automaton and the one that would prevail from the nineteenth century on, he nonetheless perceives Leibniz to be the "patron saint" of cybernetics to the extent that the philosopher is the first, in Wiener's view, to have fully understood the implications for human knowledge of the possibility of processing information using artificial machines. In his most well-known, and most lengthy, expression of praise for his German predecessor, Wiener writes:

> The philosophy of Leibniz centres about two closely related concepts—that of a universal symbolism and that of a calculus of reasoning. From these are descended the mathematical notation and the symbolic logic of the present day. Now, just as the calculus of arithmetic lends itself to a mechanisation progressing through the abacus and the desk computing machine to the ultra rapid computing machines of the present day, so the calculus ratiocinator of Leibniz contains the germs of the machina ratiocinatrix, the reasoning machine. Indeed, Leibniz himself, like his predecessor Pascal, was interested in the construction of computing machines in metal. It is, therefore, not the least surprising that some of the same intellectual impulse which has led to the development of mathematical logic has at the same time led to the ideal or actual mechanisation of processes of thought.[45]

Now, as Leibniz scholarship, Wiener's claims scarcely stand up. By Leibniz's own avowal, his philosophy centers around very different concepts than universal symbolism and the calculus of reasoning (to wit, the twin "labyrinths" of freedom and of the mathematical continuum). But Leibniz was sufficiently polymathic to ensure that virtually any characterization of his interests is not entirely incorrect. The fundamental commitment that unites Leibniz and Wiener is the idea—to use Wiener's language—that both living systems and information-processing systems operate according to the same laws, and may be studied by one and the same science; or, to use Leibniz's language, that both minds and living bodies are "automata," and as such their activities can be accurately modeled by machines, or, as we would say today, that their activities may be automated.

However, for Leibniz, a machine that *models* the activity of a living body or a mind does not *reproduce* that activity. A machine such as Vaucanson's digesting duck (invented after Leibniz's death, but certainly not beyond the bounds of conceivability during his lifetime) differs from a true duck in that the machine can be broken down in a short number of steps into parts that are not themselves machines, whereas in Leibniz's view a duck's or any other animal's body remains a machine in its least parts ad infinitum. The duck's body is an automaton nonetheless in the sense that it is causally self-sufficient; the sequence of its states is determined from within rather than as a result of outside forces pushing it around.

The human mind (and indeed, for Leibniz, the duck's mind) is similarly an automaton, to the extent that the sequence of its thoughts is determined by its own internal reasons rather than from outside. It is "running its own program" rather than being constrained from outside, yet when some of the powers of the mind are made to run on an actual machine, as happens with a reckoning engine or an analytic engine or computer, that machine, while automatic in the same way the mind is, is nonetheless only modeling or mirroring mental activity, and not engaging in mental activity. The machine reproduces the *operations* of the mind, as Leibniz is careful to put it, such as arithmetic or logical inference, but the operations of the mind by themselves neither produce consciousness nor do they require it in order to be carried out.

Wiener for his part does not find for himself an occasion to accept or reject the essence of Leibniz's mill argument. In contrast with Leibniz, he does not have any theoretical commitment to the irreducible reality of thought or perception as something of which only immaterial minds are capable. In fact, Wiener is operating in a broader scientific context and era in which commitment to immaterial minds is generally only possible as a private belief or an article of faith, not as an element of one's professional endeavors. Cybernetics is both very innovative and very much of its era. It is exclusively interested in the sort of mental operations that a machine could reproduce, without any explicit interest in the question whether in reproducing these operations the machine is doing the same thing a human mind does, or only simulating what a human mind does. Yet Wiener's quietism on this point should in no way be taken as a rejection of the basic commitments Leibniz defends with his mill argument, even if the twentieth century was much less interested than the seventeenth in preserving and defending an autonomous space for the activity of the mind on which no machine could ever encroach.

To be convinced by the idea that the human mind is a sort of reckoning engine is to suppose that what it does is fundamentally akin to what computers do when they, say, predict future consumption habits from past purchases at Target, or to what Facebook does when it flags suicide risks by detecting relevant keywords in online posts. It is to ignore, rather than to refute, the problem Leibniz identified in the mill argument. And those who do this also tend to see our computers differently than Leibniz saw his stepped reckoner: not as prostheses to which we outsource those computational activities of the mind that can be done without real thought, but as rivals or equals, as artificially generated kin, or as mutant enemies.

In this respect, though the defenders of strong AI and of various species of the computational theory of mind might accuse defenders of the mill argument and its variants of attachment to a will-o'-the-wisp, to a vestige of prescientific thinking, they are sooner the ones who follow in the footsteps of the alchemists such as Roger Bacon, and of the people who feared the alchemists and their dark conjurings. The imminent arrival of strong AI is in many respects a neo-alchemist idea, of no more real interest in our efforts to understand the promises and threats of technology than any of the other forces medieval conjurers sought to awaken, and charlatans pretended to awaken, and chiliasts warned against awakening. Technology poses plenty of real existential threats to humanity and to life on earth. Automated technology in particular poses plenty of real threats to human thriving and to political equality and justice. Lucid scientists and risk analysts will address these threats undistracted by science fictions.

These considerations may seem to have brought us somewhat far afield from our principal concern with the internet, dealing as they do with computing in general. But, as we will continue to presuppose in the following chapter as well, these two cannot be easily separated, certainly not in reality, but also not in theory. For one thing, in a very practical sense, increasingly the "training" of AI systems relies on neural networks that soak up information from the entire internet. As Alex Garland's remarkable 2014 film, *Ex Machina*, conveys—updating a timeless plot conceit from Charles Perrault's folktale *Bluebeard*—a robot that gains a sense of self, a will, and a consciousness as a result of its complex learning ability, is one that is today best imagined as learning from the totality of data floating around out there, in text messages, chatrooms, search engines, and so on. When we move back from science fiction into reality, this is also how the language-learning neural network known as GPT-3 has recently been able to master human-like language and reasoning with such uncanny perfection.

Moreover, those who maintain that human existence might well be a video game–like simulation would likely not have come to think this way if the video games in question were, say, arcade consoles featuring Pac-Man or Space Invaders. Rather, the idea that our existence is a video-game simulation is one that has emerged in a historical moment in which we are witnessing the widespread creep of the notion of "game" into all aspects of our social life. This is the case in such technical academic subfields as game theory, but more broadly it is also programmed into the web platforms that increasingly shape our conception of our own identities and structure our lived experience.

Social-media platforms like Facebook and Twitter are, in the end, video games, and so is LinkedIn, and so is ResearchGate. The social-media platform I know best, Twitter, has slowly revealed its video-game nature to me as I have become more familiar with it. Twitter is a video game in which you start as a mere "reply guy," and the goal is to work your way up to the rank of at least a "microinfluencer" by developing strategies to unlock rewards that result in increased engagement with your posts, thereby accruing to you more "points" in the form of followers. Conversely—and perhaps somewhat in their defense—Fortnite and other such massively multiplayer first-person-shooter video games are also, inter alia, social-media platforms: the kind of virtual bonds and enmities that can be forged between teenagers in such settings are no more nor less real than those forged between adults arguing about politics on a microblogging site. The programming is fundamentally the same, but with different graphics. And together, all of these platforms are contributing to the gamification of social reality, already discussed briefly in the first chapter.

What is said and thought about artificial intelligence in the present moment is a reflection of broader developments in the way specifically *networked* computing has shaped our understanding of self and society, so that even if the histories of computing and telecommunication networks are at certain moments separate, by now the way we think about deep philosophical questions concerning the ontology of machines and the possibility of machine intelligence finds these two histories intricately interwoven.

With this in mind it will be worthwhile, in the following chapter, to continue to look at some other, and certainly less familiar, dimensions of the early history of computing, where, although the theorists and inventors were not specifically focused on developing networked systems of machines, they were nonetheless interested in thinking about the natural, cosmic, and mathematical principles that structure the world as a whole, and that thereby enable a particular machine in a particular location to do its beautiful work.

4

"How closely woven the web"

THE INTERNET AS LOOM

Warp and Woof

"To strengthen our social fabric and bring the world closer together." In this seemingly straightforward description of Facebook's reason for being, which we have already considered in the introduction, Mark Zuckerberg is relying on two metaphors so subtle that they might easily pass unnoticed. Society is not actually woven, and telecommunication does not shrink physical distances between people, but only creates the appearance of proximity.

These metaphors are deeply rooted in human history, in the way human beings in different places and times have conceptualized society, nature, and society's place in nature. In the history of western philosophy, in fact, one of the most enduring ways of conceiving the connectedness of all beings (and of human beings eminently among these) has been through the idea of a "world soul": an immaterial or quasi-material principle that pervades all of nature and that causes the clouds atop the Himalayas and the plankton sunken to the bottom of the Mariana Trench, and perhaps also the cosmic dust out beyond Pluto, to be bound together in a substantial and unified whole, in just the same way that my big toe is bound together with the tip of my nose.

We do not today generally think of a human body's unity or integrality as a result of the presence of a soul, yet the metaphor proposing a fundamental similarity between the organic wholeness of the body on the one hand, and the organic wholeness of the world on the other, evidently still has quite a bit of purchase. This metaphor, moreover, is part of what guides the millennia-long drive to implement a system of telecommunication that unites people across long distances. Long before we had "proof of concept" for telecommunication networks, we had the concept itself, firing imaginations and stimulating inquiry into the principles and properties of the "ether," of conductive metals, of sound. One might dare to say, and I am in fact saying, that we always knew the internet was possible. Its appearance in the most recent era is only the latest twist in a much longer history of reflection on the connectedness and unity of all things.

Of course not everyone has believed in this unity, and some have explicitly rejected the idea of a world soul on the grounds that individual beings are quite evidently cut off from one another within their own discrete existences. Thus, in his 1659 work, *The Immortality of the Soul,* the English philosopher Henry More reasons that "if there be but one Soul in the World . . . a man cannot lash a Dog, or spur a Horse, but himself would feel the smart of it."[1]

More's critique of the idea of the interconnectedness of all beings (or at least of sentient ones) finds a peculiar antithesis in the commonplace observation we often hear today about the power of mass media to sensitize us to the suffering of others. In the most recent wave of protests against police brutality in the United States in 2020, one of the most common remarks on

Twitter asks us, as a certain user known only as "john" put it (for which he quickly received more than 85,000 likes), "how much cops would get away with if iphones weren't a thing nowadays."[2] We plug ourselves into the flow of images and it changes our brain chemistry: the kittens and babies trigger dopamine production, and the police brutality triggers cortisol. The two sorts of image are carefully balanced by the owners of the media through which the images are transmitted, in the exclusive interest of maximizing their own profits. Yet the balance between the kitten's playful joy and the police victim's sorrow is also a fairly accurate reflection of the nature of reality, which, like the social-media feed, consists in both joy and sorrow, wonderful love and gross injustice. In transmitting the world, and making us feel it, or feel as if we feel it, the internet also gives us a handy, reduced, caricatural but not alien, microcosmic mirror of our world.

I am not, here, going quite so far as to say that the internet proves the truth of the theory of the world soul as it descends from Greek antiquity to the present day. I am too responsible to say that. Rather, I will carefully venture, as I began to do in the previous chapters, to note that it will help us to understand the nature and significance of the internet to consider it as only the most recent chapter in a much longer, and much deeper, history.

The other metaphor evoked by Zuckerberg speaks of the social "fabric." There is perhaps no more common image than that of the weaver or of the loom in the way thinkers throughout history have sought to capture the nature of the interconnection of things and of people. And indeed this metaphor commonly occurs in support of arguments in favor of a world soul. Thus, as already briefly mentioned in chapter 2, the second-century Roman statesman and Stoic philosopher Marcus Aurelius implores his readers:

> Cease not to think of the Universe as one living Being, possessed of a single Substance and a single Soul; and how all things trace back to its single sentience; and how it does all things by a single impulse; and how all existing things are joint causes of all things that come into existence; and how intertwined in the fabric is the thread and how closely woven the web.[3]

What does it mean to say that all of creation is "woven," and that therefore all creatures are "interwoven" with one another? To consider another example widely distant in place and time, the *Brihadāranyaka Upanishad,* composed in India in the seventh century BCE, describes our world as being "woven in water, like warp and woof," while the water is woven in air, the air woven in "the worlds of the sky," and so on.[4] This enumeration of instances of cosmic weaving is unique, and uniquely beautiful, but the image of weaving itself is not at all uncommon in ancient cosmological writing throughout the world.

When we move away from myth and poetry and back into the history of technology, moreover, we find weaving machines not only developing in parallel to information-processing machines from the beginning of the modern period going forward. We find, much more, that the history of looms and the history of computers is at certain moments *literally* one and the same history, as we shall now see.[5]

Algebraic Weaving

In 1808, the French inventor Joseph Marie Jacquard introduced to the world his automated loom, capable of transferring a design onto silk that had been "programmed" into a sequence of punched cards.[6] At first glance it might not seem that the punched-card weaving machine deserves a place in the history of computer science, alongside other technologies more narrowly focused on data processing rather than on the manufacture of a product. Yet consider: today some of the most remarkable innovations in computing are taking place in 3D printing and in the development of the so-called internet of things, or of physical objects networked together by sensors and software, "smart homes," smart energy grids, remote health monitoring, shipping, and so on, all of which trace their origins back at least as much to the manufacturing machines of the industrial era as they do to the reckoning machines and analytical engines of the same period.

The automation of silk production, a process in which Jacquard's loom is a relatively late development, involves a long history of inventors trying to understand and master what today we would call the "interface" between living and artificial systems. One of Jacquard's predecessors in the development of loom technology, Jacques de Vaucanson, who himself experimented with punched cards as early as 1725 but did not employ them with any significant degree of automation, is much better known for his so-called *canard digérateur* or "digesting duck," which we have already mentioned in the preceding chapter: a mechanical waterfowl so lifelike, its inventor promised, that it could not only walk around and flap its wings, but could even consume food and defecate its waste products.[7] It is not surprising that one and the same man should have been occupied with these two projects, for both machines, the defecating duck and the silk-weaving loom, are doing the same sort of work, though in opposite directions: the duck starts with artifice and seeks to push it across the border that separates it from the natural; the loom starts with a natural product, the silk of certain species of moth larva, and turns it into the artifice of a woven piece of fabric.

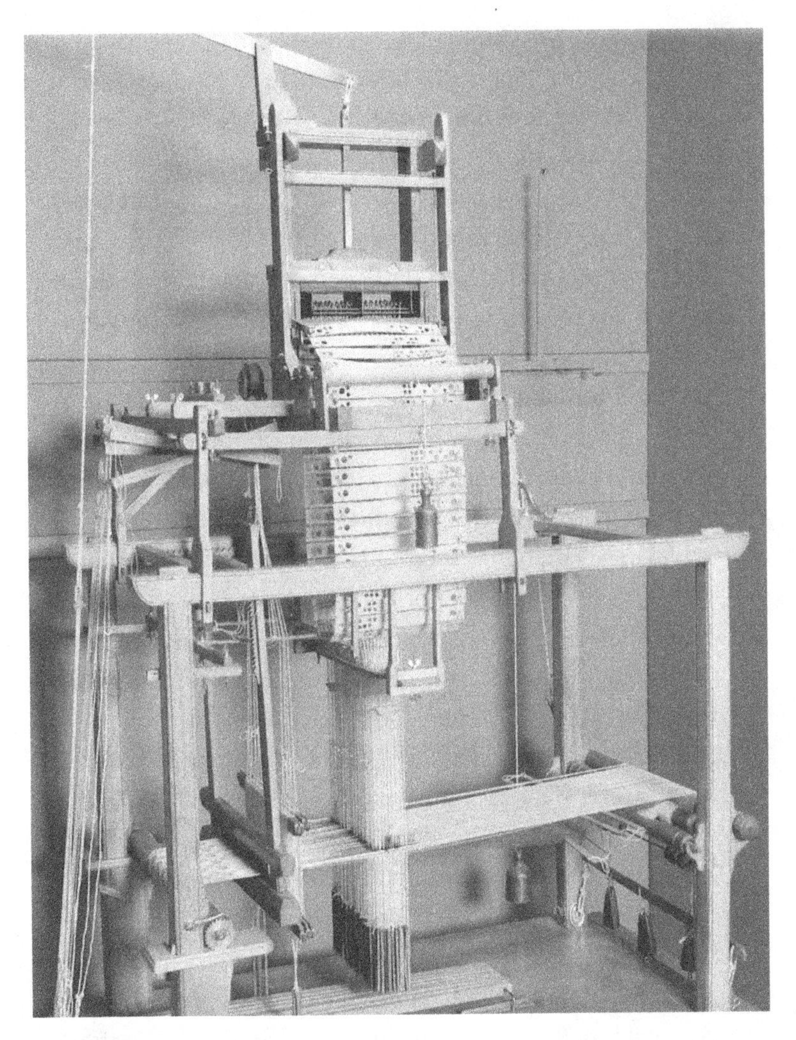

FIGURE 5. Joseph Marie Jacquard's punched-card loom, 1808. Musée des Arts et Métiers, Paris.

Perhaps no one in modern history has been more interested in exploring the boundary between the artificial and the natural than G. W. Leibniz—no stranger to the reader by now. Like Vaucanson, Leibniz was particularly interested in understanding just how far the analogy between animal bodies and machines could be pushed. And, like Vaucanson, Leibniz also had a special interest in silk: he developed an elaborate plan to fund a scientific society with revenues from the patent he obtained in 1703 for silk manufacture in Berlin.[8]

Wherever there have been people weaving silk, it is safe to say, there have been people thinking about the way in which human artifice supervenes on the natural world, and about the power of the metaphor of weaving for conveying something of the way in which the things and beings of the social and natural worlds are woven together with one another, even across long distances. Leibniz's closest successor in this cluster of preoccupations is surely Ada Augusta Lovelace, the British mathematician and daughter of the romantic poet Lord Byron, who conceived her own work as a variety of "poetical science." In addition to being an illustrious woman of what is today known under the impoverished label of "STEM," Lovelace also provided, more clearly than any of her contemporaries, an explicit account of the overlap between the history of weaving and the history of computing.

Born in 1815, while still a teenager in the early 1830s Lovelace began a long collaboration with the mathematician Charles Babbage, who is best known for his work on the so-called Analytical Engine. Prior to meeting Lovelace, Babbage had by 1822 already designed a "Difference Engine," a mechanical calculator that advanced beyond the early modern reckoning engines of Leibniz and others in that its primary use was in tabulating polynomial functions. The later Analytical Engine, by contrast, was to be a computer in the full and proper sense, capable not only of executing computations, but also of instantiating algorithms by means of conditional branching and looping, and of information storage.

Prior to his work with Lovelace, Babbage had conceived of the possibilities for his engines as exhausted by operations upon numbers. Like Leibniz, Babbage thought that the greatest potential for our use of calculating machines lay in their use as prosthetics of the human mind, doing for us what is too time-consuming and drudging to be worthwhile, yet still leaving to us, and us alone, those operations of the mind that machines cannot possibly reproduce. While the respective contributions of both collaborators have been the subject of much scholarly debate, it is clear enough that Lovelace was the member of the pair who best understood the full potential, beyond mere number-crunching, of machines as extensions of human thought.

This much becomes clear in the description of Babbage's Analytical Engine published by the Italian mathematician Luigi Federico Menabrea in 1842. Babbage had participated in a conference of the Turin Academy of Sciences two years earlier, where the two men met. Although Menabrea was destined for a career as a statesman, eventually becoming prime minister of Italy in 1867, his meeting with Babbage was decisive for both men's futures. It was thanks to Menabrea's sketch of Babbage's work that the Analytical Engine would come to have the influence it did, and Menabrea for his part continued to pursue scientific projects throughout his political career that were in some way related to machine-based analysis.

The Italian scientist describes the significance of Babbage's Analytical Engine as lying in the fact that it transfers "the mechanical branch" of mathematics, fittingly, to a machine, while "reserving for pure intellect that which depends on the reasoning faculties."[9] This is a clear echo of Leibniz's hope for his own reckoning engine almost two centuries earlier, that it would liberate human minds from the time-consuming task of tediously performing their own calculations, thus freeing them up for the investigation of more fundamental mathematical and philosophical problems. There is nothing inherently ennobling, for Babbage as for Leibniz, about doing long division or other arithmetical calculations that a machine can do just as well (any student of mathematics who has ever been told that calculators are not allowed during a test might thus attempt to appeal this prohibition by invoking the names of these mathematical giants, and by noting that for them the prosthetic work done by a calculating machine is not the sign of a weak mind, but of a mind that simply has more important things with which to occupy itself).

Babbage's new machine, Menabrea writes, "can of itself, and without having recourse to the hand of man, assume the successive dispositions suited to [its] operations," in contrast to the earlier difference engine, which constantly had to be reset manually for each new operation. Menabrea does not conceal the true origin of this new innovation: the solution of the problem of automated operation, he writes, "has been taken from Jacquard's apparatus."[10] Jacquard, recall, is the inventor of the loom that uses punched cards for the production of images in woven silk. This machine, which would ultimately decimate the traditional silk manufacturers by drastically lowering the labor costs in the production of its fabrics, may rightly be seen as an early blow in the wave of such "disruptions" that began to transform manufacturing in the industrial revolution, and that today, in the midst of the information revolution, have now moved on to disrupt and deform the work of writers, artists, and thinkers. The industrial revolution and the information revolution, in other words, began at the same time, if we take Menabrea at his word and we see Jacquard's loom as lying at the beginning of both of these revolutions. The principal difference is that the destruction of the applied arts and crafts by industrialization began to make its effects felt generations before the (currently ongoing) destruction of the fine arts and the belles lettres by the automation of information-processing and "content" production.

Menabrea explains that formerly the process of production of silk garments "was lengthy and difficult, and it was requisite that the workman, by attending to the design which he was to copy, should himself regulate the movements the threads were to take."[11] Thus, "brocaded stuffs" remained very expensive and rare. But Jacquard's loom was able to connect all of the threads that were to be used, "with a distinct lever belonging exclusively to that group."[12] All these levers, he explains at some length, "terminate in rods, which are united together in one bundle, having usually the form of a parallelopiped with a rectangular base. The rods are cylindrical, and are separated from each other by small intervals. The process of raising the threads is thus resolved into that of moving these various lever-arms in the requisite order."[13] This is done, in turn, by the use of "a rectangular sheet of pasteboard," by which Menabrea has in mind, precisely, the punched card—the clavis of all automation and information-processing technology to come—which renders a "text" to be "read" in a "language" that even a machine can understand, namely, a pattern of presences (the physical barrier of the card) and absences (the holes punched in the card). Presences and absences may also of course be represented as zeroes and ones, which Leibniz as well had already understood, even if he did not come up with a mechanism for reading these.

For someone interested primarily in the operations of Babbage's information-processing machine, Menabrea goes into surprising detail in his account of the operations of Jacquard's weaving machine. He distinguishes among other things between the limited utility of a solid card, one that is as it were "all presence," and a punched card or pasteboard that alternates presence and absence, solidity and holes, ones and zeroes. If the former sort of card is used, Menabrea explains, "and an advancing motion be then communicated to the pasteboard, this latter

According to Lovelace, Babbage's earlier Difference Engine had been strictly arithmetical, "and the results it can arrive at lie within a very clearly defined and restricted range, while there is no finite line of demarcation which limits the powers of the Analytical Engine. These powers are co-extensive with our knowledge of the laws of analysis itself, and need be bounded only by our acquaintance with the latter."[18] Unlike Babbage or Menabrea, but very much like Leibniz before her, Lovelace wishes explicitly to establish an analogy between information-processing and world creation: an Analytical Engine is capable of mirroring the actual world and indeed of generating infinitely many possible worlds besides. In this respect, the computer, alone among human inventions, Lovelace thinks, brings within our reach the power to know the mysteries of God's creation: to simulate the world on a computer brings with it the hope of arriving at "maker's knowledge" of the world—again, to invoke Francis Bacon's helpful phrase—and thus at knowledge of the world as deep and thorough as any other knowledge of the objects of our own manufacture.

The science of computing, Lovelace believes,

> constitutes the language through which alone we can adequately express the great facts of the natural world, and those unceasing changes of mutual relationship which, visibly or invisibly, consciously or unconsciously to our immediate physical perceptions, are interminably going on in the agencies of the creation we live amidst: those who thus think on the mathematical truth as the instrument through which the weak mind of man can most effectually read his Creator's works, will regard with especial interest all that can tend to facilitate the translation of its principles into explicit practical forms.[19]

After this poetic account of the significance of her own life's work, Lovelace adds a surprising reflection. Going far beyond what Menabrea himself had acknowledged of the debt of Babbage's Analytical Engine to Jacquard's loom, she explicitly claims that the engine really is just another sort of weaving machine: "The distinctive character of the Analytical Engine," she writes, "is the introduction into it of the principle which Jacquard devised for regulating, by means of punched cards, the most complicated patterns in the fabrication of brocaded stuffs . . . We may say most aptly that the Analytical Engine *weaves algebraical patterns* just as the Jacquard-loom weaves flowers and leaves."[20]

Lovelace believes that as a result of the use of the punched cards, the Analytical Engine is of a wholly different character than the Difference Engine that preceded it, and indeed than such earlier devices as the Leibnizian stepped reckoner or the Pascalian Pascaline. Those machines only computed, she believes, while this one is capable of a full and rich simulation of human thought; those machines only reckoned, to use Cantwell Smith's terminology introduced in the previous chapter, while this one is capable at least of simulating *judgment*. The Analytical Engine, Lovelace believes, "does not occupy common ground with mere 'calculating machines.' It holds a position wholly its own; and the considerations it suggests are most interesting in their nature. In enabling mechanisms to combine together *general* symbols, in successions of unlimited variety and extent, a uniting link is established between the operations of matter and the abstract mental processes of the *most abstract* branch of mathematics."[21] Through the combination of general symbols, rather than of mere numbers, the Analytical Engine reproduces the most abstract mental processes involved in mathematics, particularly in algebra or in any other field in which variables operate alongside numbers.

Lovelace is not modest about the device she helped to devise. Through it, she believes, "not only the mental and the material, but the theoretical and the practical in the mathematical world, are brought into more intimate and effective connexion with each other. We are not aware of its being on record that anything partaking in the nature of what is so well designated the *Analytical* Engine has been hitherto proposed, or even thought of, as a practical possibility, any more than the idea of a thinking or of a reasoning machine."[22] What is certainly new, or at least greatly intensified, in Lovelace's era is the habit of claiming absolute novelty for what are in fact an era's gradual innovations upon projects inherited from the past. In truth, reasoning machines had been proposed, some merely fantastical, some with clear instructions for implementation, over the course of several centuries. Leibniz himself gave clear instructions for the implementation only of a calculating machine as opposed to a full-fledged thinking machine. But his own binary system entails that any thought can in principle be represented in a code that is readable in the same way that a calculating machine "reads" numbers. And before Leibniz there were numerous other inventors, alchemists, tinkerers and fanciful dreamers alike, working away in their ateliers, or perhaps only in their own minds, trying to figure out how human thought might be transferred to an external system. Try as we might to find the first parents, this is a dream with no father, or mother, and no definite origin in time.

The very development of the binary calculus that, on a certain understanding, marks the true beginning of the history of information science, was itself a direct borrowing from a broadly neo-Platonic mystical tradition of contemplating the relationship between being and non-being: where the former might be represented by "1" and the latter by "0." In this tradition all of creation is itself akin to a mixture of zeroes and ones, composed together by the Demiurge in order to yield the qualitative variety of all the things of nature, which share in divine being to some extent, but, because they also have non-being mixed into them, are nonetheless not themselves divine. Leibniz understood that the way this tradition conceptualized the world itself could be adapted and applied to the treatment of language: just as the creatures of this world, the planets and stars and plants and animals, are mixtures of being and non-being, so too can we say all that might be said about the world by means of mixtures of zeroes and ones. Thus the binary calculus is born, and thus the history of modern computer science begins.

Leibniz writes to Joachim Bouvet in 1701, in the course of a comparison with the neo-Confucian uses of *I Ching* hexagrams, a topic he understood poorly but took to have much in common with his own binary system: "The new [binary] numerical calculus that I have invented . . . opens up a large field for new theorems. And above all this calculus gives an admirable representation of Creation. For following this method all numbers are written by a mixture of the unity and the zero, somewhat as all creatures come only from God and from nothing . . . Nor is it said in vain that essences are like numbers, and that all the imperfections in things consist only in negations."[23]

The mythical-cosmogonical framework of binary code, if not wholly forgotten by the early nineteenth century, is one that an inventor or a theorist could by now engage with or ignore at will, according to preference. As we have seen, Lovelace is sharply aware of it, and is happy to invoke the metaphor of creation in explaining what the Analytical Engine does.

One of the most surprising lessons of Lovelace's and Menabrea's reflections on computing is that they not only deepen chronologically, but also reverse the chronological and conceptual order of what we often take to be much more recent developments in computer science. We typically suppose, namely, that what is erroneously called "3D printing," or what is somewhat less erroneously called "the internet of things"—in short, wherever we have an information-processing machine directing its information in a way that brings about a change in the physical world, a plastic gun appearing from a mold, say, or a stove caused from a distance to commence its self-cleaning—is only the very latest development in a much longer history of computer science. 3D printing is thought to come after paper printing; the internet of things is supposed to follow the internet of screens. In fact, however, as we have been seeing, a prime candidate for the distinction of "first computer" had as its sole purpose what we may rightly call the 3D printing of brocaded stuffs. The dream of automata that transform the world is at least as old as the dream of automata that simulate the world. In the early nineteenth century an automaton of the former sort showed the way forward for automata of the latter sort.

Woven fabrics have always been a powerful metaphor both for the order of nature and for our own efforts to simulate or reproduce that order in symbolic systems. But weaving is not only a metaphor. It is also something people actually do (mostly women, in fact), one of the most primordial forms of making, which is to say, of transformation of nature into artifact by a given technique. That a new weaving technique should eventually trigger both the industrial and the digital revolutions seems surprising to us now, but should not be, given how fundamental weaving has been throughout human history.

Why Do Metaphors Matter?

We have been returning again and again over the last chapters to the problem and the potential of metaphorical language: where its boundaries are, and what it has to offer us in our effort to understand the actual world in which we live. Machines, plants, animals, humans, the world as a whole: all seem to have something real in common, which it would be useful and important to investigate, yet in our efforts to do so we often feel as though we are lapsing haphazardly into mere poetic language, and thus obscuring the commonalities rather than clarifying them. Some philosophers, aware of the constant threat of such a lapse, prefer to stop exploring the commonalities altogether, and even, against all evidence, to deny that there are any. Evidently, this book is not for, or by, that kind of philosopher.

It seems that even when we move away from the poetic invocations of weaving in Marcus Aurelius or in the *Upanishads* and in the direction of the concrete facts about punched-card technology, to dwell for too long on what the loom and the analytical engine have in common is to begin to make the literal metaphorical again. The Jacquard loom might be literally a kind of computer, but the Analytical Engine is not literally a loom, and it does not literally weave data or algebraic patterns. To hold these two sorts of thing in relation to one another, to turn around them as it were and to see the one through the other in succession, the loom through the engine and the engine through the loom, is to move between two registers of language that philosophy and science ordinarily seek to keep separate.

In fact, however, the history of science is often largely a history of metaphors. What is discovered as a new explanation or theoretical account of how the world works is, often, a new way to "carry over" (the etymological meaning of the word "metaphor") from one domain habits or even fashions of thinking and understanding into another domain where these habits and fashions were not originally intended to go. Thus early evolutionary theorists, to note one example, extend the momentum of a few centuries of thinking about human history as a progressive development beyond the social realm and into nature. Or one might, with perhaps a little more effort, work one's way from the second law of thermodynamics or the theory of relativity back to developments in nineteenth- and twentieth-century politics and culture. This is not to say that any of these theories is not true, but only that there is always a cultural history that shapes the way we come by and articulate the truth. And typically this cultural history involves the spread of concepts or tropes from one domain into another, which is, again, nothing other than a carrying over or extension of metaphor—as Norbert Wiener himself noted, for example, in the nineteenth century the study of animal physiology was overwhelmingly occupied with questions of heat transfer and similar ideas carried over from the study of thermodynamics.

There is in fact a bidirectional circulation of concepts and tropes that extends very far back in the parallel history of the biological and mechanical sciences, in the study of living systems and the study of machines. In our own day we have witnessed the extension of the concept of natural selection from the genetic level of analysis, of interest in the study of biological evolution, to the "memetic" level, of interest in the study of communication networks. The fact that the idea of the meme has caught on so succesfully is itself an instance of memetic selection.

Viruses are another example of the cross-fertility between information science and life science. The current global pandemic is in the course of revolutionizing our language in profound ways. One of the most significant changes however is not so much a revolution as a restoration: at just the moment in history when a derived and secondary sense of "virus" (as in "computer virus," "viral meme," "viral celebrity," "to go viral") was poised to move into first place, a real virus, a harmful biological agent, came roaring into history to remind us of its enduring primacy. This is an irony that future historians may be better positioned to savor, but for now we might at least benefit from a consideration of how these two notions came to be paired in our metaphorical imagination, and what this pairing might mean as we move forward in a world fundamentally shaped by these two forces: the pandemic and the internet.

It might be surprising to learn that biological viruses were not part of our conceptual and linguistic reality for very long before they made the leap (metaphorically) to computers. Viruses were only discovered, by Dmitry Ivanovsky, in 1892, though the term was used for more than a century before that to describe any agent of infection. Less than a century after the discovery of biological viruses, in 1985 Fred Cohen adapted the term to describe self-replicating computer programs that insert their code into other programs. It would not be until the age of social media, however, that this new notion of virality would take on a generally positive connotation through its combination with the related notion of the meme. And by 2020, a generation of new-media users had grown up vaguely thinking of pandemics as part of a backwards past we had long left behind, while at the same time doing whatever they could to insert themselves "virally," via their own favored fragments of culture, into the bloodstream of the age. This is, broadly, the condition in which the SARS-CoV-2 virus found us when it made its entry into human history.

In the mid-twentieth century, Wiener, along with a number of like-minded contemporaries, conceived the new science of cybernetics as pertaining equally to living systems and to information-processing systems, which had in common, in his view, the principle of feedback looping or circular causality. Wiener united in a single research project kinds of being and domains of research that are generally seen as sharing in the same nature only in a metaphorical way. But he took what they have in common literally, and in so doing revolutionized the sciences of both.

In the history of thinking about living beings and machines, it is often very difficult to say whether a given author is pushing a metaphorical point, or giving what that author believes to be a correct account of the nature of the one or the other sort of system. When René Descartes describes animals as "machines," does he mean that they are *like* machines? Or that they may best be understood by thinking about them in relation to machines? Or that there is literally no important difference between a deer and a clock? When Giovanni Borelli fashions the muscles of an animal body as so many pulleys, levers, and hydraulic pumps, does he mean to reveal to us the true nature of what he sees, or does he mean to invite us to exercise our imaginations by means of visual metaphor?

Sometimes the answers to such questions are clear, other times they are less so. But in the aggregate, these examples show us that the history of the life sciences and the history of mechanical science, and even more so of information science, are so thoroughly intertwined in their cultural reverberations as to end up perpetually influencing one another's content and direction of development. This intertwining, moreover, plainly has something to do with what the different domains of reality truly have in common, whether this commonality is expressed in terms of metaphor—as is likely the case with Borelli's hydraulic dog muscles—or in terms of scientific explanation—as is the case in the cybernetic research program as it developed in the second half of the twentieth century.

A metaphor, again, is literally a carrying-over of a concept from one semantic domain into another where it is typically thought not to belong. But when this carrying-over endures and recurs, imposing itself as if spontaneously in different times and places, we might ask ourselves whether it is not so much being smuggled into an alien territory as it is striving to return home.

A related example might help us to understand this point more clearly. Throughout most of human history in most cultures, non-human animals were seen as "persons," that is, as individual actors in a sociocosmic reality that included both the social and the natural realms. To the extent that they were persons, it was not at all problematic to attribute to them "personalities." In modern Europe beginning in the seventeenth century, and with a stark radicalization in the nineteenth and twentieth centuries, this common belief in the personhood of animals was displaced by a new, ideologically motivated orthodoxy, which saw only human beings as persons, and animals as things, machines, or in some way as imitations of ensouled beings. Under this orthodoxy, the attribution of personalities to animals was dismissed as "anthropomorphism" and as naive metaphorical extension of human traits to a category of entities that cannot share in these traits.

And yet, it just kept happening: people kept right on defying the orthodoxy and describing animals as they always had. The mechanical model of animal physiology defended by Descartes, and the behaviorist model of animal psychology later adopted by B. F. Skinner and others, have now largely been displaced by a more nuanced scientific view, according to which animals have complex inner emotional lives just like human beings. Many claims that would have been judged as anthropomorphizing metaphor a century ago are now held to be literal descriptions.

Again, this is only an example meant to illustrate the mobility of the boundary between metaphor and literal description. Sometimes, if the folk display a persistent habit of speaking in a way that official doctrine deems metaphorical, but that seems truthful to the speakers, the problem might lie with official doctrine, and not with the folk. (We must emphasize "sometimes" here: an appeal to popular habits of speech in general is never in itself a decisive answer to difficult philosophical questions.)

No two different things are *entirely* different, and sometimes it makes sense to assert that a feature of the one that resembles a feature of the other in fact does so because these two things share, to some extent, in the same nature. But when are we in the presence of such a case? And how do we know? An analogy, as distinct from a metaphor, is suitably invoked when two things have similar features, and when these features are the result of the same natural forces or laws, even though the two things do not necessarily share in the same nature.

It is with just such an understanding of analogy that Kant is working in the *Critique of the Power of Judgment* when, as we already saw, he defines it as the identity of the relation between the causes and effects of two distinct things, notwithstanding the specific difference between the things.[24] Kant is interested in the sort of analogy in which we understand the cause or ground of only one of the two sorts of thing placed in analogical relation, as in the artistic productions of human beings (the cause of which we do know) and the appearance of beautiful forms in nature (the cause of which we do not know). But consider in turn an example in which both sorts of thing placed in analogical relation to one another are roughly equally well known to us: the whorls in a snail's shell and the whorls in a galaxy are truly analogous—they instantiate the same mathematical proportions as a result of their presence in the same universe governed by the same general physical laws—even if for most purposes we would not say that these two things share in the same nature. Nothing else follows about gastropods and galaxies from the fact that they both furnish an example of the Fibonacci sequence, and yet it is meaningful and true to point out this analogy of structure.

An example of a metaphor by contrast, again understood literally as a carrying-over of a property into another domain where it is usually held not to belong, would be to say that a galaxy is "slimy," or the little balls at the end of a snail's antennae (in fact its eyes) are its "twinned stars." This much should be clear enough, but only because we already know enough about star systems and gastropods to know when the purported similarity is well-founded and when it is being carried over by our imagination.

Difficulty arises when we are still in the course of inquiring into the nature of a given thing—as we are doing here regarding the nature of the internet—for it assumes too much at the outset to suppose that we already know what can or cannot be carried over in the course of an explanation. It assumes too much, that is, to suppose that we already have a clear understanding of where the boundary between analogy and metaphor lies. And this is precisely our predicament when we attempt to comprehend computers, networked and otherwise, in their relationship to other things: to living nature, to human minds, or to looms.

Threads

The Analytical Engine was not, and was never conceived to be, a networked device. It is thus principally an ancestor of what would in the twentieth century become the pocket calculator or the word processor, and not of what would become the internet. Babbage's and Lovelace's engines were meant to sit in one place and to process the information fed into them in that same place, not to absorb information from far and wide.

A mostly separate ancestral lineage of the internet would witness the successive development of systems and devices that were not meant to process information, but only to transmit signals: the lunar listening disc, the electric telegraph, the snail telegraph, the recording sponge, the optical semaphore, all of which we have considered in previous chapters (and some of which have no existence beyond the human imagination). Yet we must emphasize "mostly," for it is clear that part of the history of computing involves the idea that no information-processing system is fully closed, that wherever ideas are being processed, they are somehow "in the ether." In the earlier and more mystically informed investigations of thinkers from Llull to Leibniz, as we discussed briefly in our treatment of Semyon Korsakov, ideas themselves had not yet been fully sucked out of the world itself and into the minds of individual people. Along a similar line, for a machine to process ideas, for example the ideas of "substance," "being," or "God," is for it to connect with the objective order of the world, to be "networked" as it were with reality as a whole.

It is partly in light of this legacy, the way the line between the computing engine and the telecommunication device gets blurred the further back and the deeper we go, that the metaphors of loom and thread turn out to be so compelling, and the literal historical fact of the debt the computer technology has to weaving technology reveals its full significance. For as we have seen, the very raw material that is woven in literal looms—the thread, the filament—is also the most commonly hypothesized entity, or principle, or something between an entity or a principle, in efforts from antiquity to the modern era to account for what may generically be called "action at a distance": all the different ways in which actions or events in one place appear to bring about effects in another place, without any apparent contact or obvious mediation of physical particles.

Thus, for example, in 1661 the Jesuit scientist Franciscus Linus maintained that there is a *funiculus*, or "little rope," too small to be seen, that holds the mercury up inside a barometer (he failed to grasp that it is the same thing that the instrument is measuring in the outside world—air pressure—that also causes the fluid within the instrument of measurement to behave the way it does).[25] Or in a 1679 textbook of Cartesian physics, Antoine Le Grand explains that an odor "consists in innumerable filaments" moving through the air, of a size that can be detected only by the nose but not by the eyes.[26]

And of course the figure of the thread has also made its way into the literal language of contemporary theoretical physics, or at least the language that is as literal as can be in such a speculative domain. At both the smallest and the largest scales of physical reality, there are said to be entities for which "thread" would be a close synonym. At the microscale, some theorists postulate the existence of "strings," held, if they exist at all, to be no more than 10^{-33} centimeters long. At the extreme opposite end of the scale, physical cosmologists have identified what they call "galaxy filaments," the largest known structures in the universe, consisting of several superclusters of gravitationally bound galaxies.

Scientists license such metaphors, and science journalists run with them, shaping the way ordinary people understand the world around them. Thus a recent newspaper article describes galaxy clusters as being "connected by spidery filaments in what's known as the cosmic web."[27] From the smallest to the largest scales, the stuff of this world is still said to be bound together by threads. The metaphor holds the fabric of our explanations together, at least when we move out of the pure mathematics that is in important respects the true language of theoretical physics, and we struggle to offer up natural-language descriptions of the world that remain largely faithful to the numbers.

As Paul Ricoeur has argued, metaphor arises "from the very structures of the mind,"[28] and in this respect is at least as worthy of being taken seriously by philosophers as any literal proposition. This is particularly so when the human mind keeps returning inexorably to the *same* metaphors in accounting for some difficult or intangible aspect of the structure of the natural world, as for example the idea that nature or the cosmos is built up out of strings or threads, or that information may be woven on a computer just as threads are woven on a loom. The brute historical fact of the intersection of the histories of weaving and computing in Jacquard's invention is not so much a literal exception to an otherwise metaphorical trope as it is a reminder of the strength of metaphor itself: sometimes the structures of the mind are powerful enough to pour out of the mind, and to impose themselves in our built reality as well. Manners of speaking become manners of world-building.

5

A Window on the World

I hear new news every day, and those ordinary rumours of war, plagues, fires, inundations, thefts, murders, massacres, meteors, comets, spectrums, prodigies, apparitions, of towns taken, cities besieged in France, Germany, Turkey, Persia, Poland, &c., daily musters and preparations, and such like, which these tempestuous times afford, battles fought, so many men slain, monomachies, shipwrecks, piracies and sea-fights; peace, leagues, stratagems, and fresh alarms. A vast confusion of vows, wishes, actions, edicts, petitions, lawsuits, pleas, laws, proclamations, complaints, grievances are daily brought to our ears. New books every day, pamphlets, corantoes, stories, whole catalogues of volumes of all sorts, new paradoxes, opinions, schisms, heresies, controversies in philosophy, religion, &c. Now come tidings of weddings, maskings, mummeries, entertainments, jubilees, embassies, tilts and tournaments, trophies, triumphs, revels, sports, plays: then again, as in a new shifted scene, treasons, cheating tricks, robberies, enormous villainies in all kinds, funerals, burials, deaths of princes, new discoveries, expeditions, now comical, then tragical matters. Today we hear of new lords and officers created, tomorrow of some great men deposed, and then again of fresh honours conferred; one is let loose, another imprisoned; one purchaseth, another breaketh: he thrives, his neighbour turns bankrupt; now plenty, then again dearth and famine; one runs, another rides, wrangles, laughs, weeps, &c. This I daily hear, and such like, both private and public news, amidst the gallantry and misery of the world.

—ROBERT BURTON, *THE ANATOMY OF MELANCHOLY*

Unconfined Thoughts

I am writing, from New York City, during the coronavirus quarantine in the spring of the year 2020. I have in more recent weeks begun to venture out for short walks, though for more than two months I did not leave our cramped one-bedroom apartment in Fort Greene, Brooklyn. The New York Public Library, where I have a fellowship for a year of research, remains closed, and I am painfully cut off from my precious books. I think of Erich Auerbach, who wrote his magisterial *Mimesis: The Representation of Reality in Western Literature*, while in Istanbul in exile from Nazi Germany.[1] He, too, was cut off from his books, and as a result his work in exile stands as a sort of monument of making do, an effort to create the best scholarship he could under difficult conditions, but also as a book *about* the absence of books, about what it is like to represent books in memory, and to meditate on the power these absent books have of representing realities that are absent within them in turn.

In many fundamental respects the present project must of course not be compared to Auerbach's. My circumstances are not nearly as trying. For one thing, I have the internet.

Yet if I recall Auerbach's work, it is for the simple reason that here, too, I find myself writing a book whose writing is made difficult by the absence of books, but that in turn becomes at least in part a book about this absence. There are far more endnotes citing websites than I have ever permitted myself in my previous books. And while we may attempt to write this off as temporary compensation, the truth is that the pandemic has really only pushed us over a ledge on which we already teetered. In so many ways, there is no going back. This book is both impacted by the present crisis, but is also to some extent an attempt to document it, to make sense of it, and to discern what may come next.

Like so many other people, I have taken to Zoom and to other internet platforms to seek out whatever human contact I can with so many people I know and love. Some of the people I have found on the other side of my screen, especially the younger ones, have truly been suffering from their new confinement. Some, more like me in age and in temperament, have confessed that they had been waiting their whole lives for this moment, and barely jokingly have proclaimed their hope that the lockdown should never end. It is not that they are prepared to relinquish the world, exactly, or to enter a monastic condition and to silently contemplate only their own mortality and finitude. The dream that is now being realized is to retreat not into a closed cell, but a cell from which to look out, as through a magic window, on the totality of the dazzling world from which we have been physically cut off.

While the internet, with its browser windows that give us access to so much beyond the bounds of our confinement, has fundamentally defined the quality of our present quarantine experience, the dream that I and other like-minded people are now living is one that was also pursued and relished in centuries past. If I could nominate a patron saint for the people I have identified, and with whom I identify—the people we might perhaps call "worldly confiners," in contrast with the world-renouncing confiners in the monasteries and convents—it would surely be Robert Burton, the author of the strange and unclassifiable 1621 treatise, *The Anatomy of Melancholy*. Nominally a work of humoral medicine, investigating the causes, symptoms, and remedies of the surfeit of black bile in the body, the author takes the opportunity authorship affords to deliver up a sort of commonplace book, a compendium of trivia, a confessional memoir, a simultaneously satirical and heartfelt encyclopedia of everything that mattered to its creator. The novelist and philosopher William Gass has written movingly of Burton's life work that it reveals "the width of the world that can be seen from one college window . . . ; what a love of all can be felt by one who has lived it sitting in a chair."[2]

Burton himself, though he learns so much daily of the affairs of his fellow beings, confesses: "I live still a collegiate student, as Democritus in his garden, and lead a monastic life . . . sequestered from those tumults and troubles of the world."[3] He says that he has "never travelled but in map or card, in which my unconfined thoughts have freely expatiated, as having ever been especially delighted with the study of cosmography."[4] And he describes himself as a "mere spectator of other men's fortunes and adventures, and how they act their parts," and imagines that these are "diversely presented unto me, as from a common theatre or scene."[5]

The image of the world as a stage or theater, of course also familiar from Shakespeare, is one that ordinarily takes for granted that we already have a clear idea of the literal sense of "world" that is being placed into a metaphorical relationship with "stage." But what is a world? Is it simply the sum of all physical objects in the universe? Is every star the center of a world, as early modern authors like Blaise Pascal and Henry

More tended to suppose, with its own beings living from its light and warmth as we do from our sun? Or is the world something more "mundane," as in the social reality implied by the title of the leading French newspaper, *Le Monde*? Are there as many "worlds" as there are social milieux: a world of Broadway theater people, a world of pediatricians, and so on?

The lost art and science of cosmography, to which Burton alludes, was often capacious enough to include all of these levels of focus under the notion of "cosmos," which might be rightly understood as the closest Greek equivalent to the Latin *mundus* or to the English "world." As established paradigmatically in Sebastian Münster's 1544 *Cosmographia*, a cosmos or world is any ordered whole, as opposed to the potentially infinite and disordered expanse of the universe. A cosmos or world is what matters. What gets counted as *the* world for a person will always be the product of the individual's particular idiosyncratic and subjective encounter with society, nature, and so on. To say that Burton "loved the world" is to say that he sought to include as much as he could on his imaginary stage, to give it order, and to allow it to matter to him. The result is a deeply personal, subjective, and often histrionic approach to science, whether of the human body, or the celestial bodies, or of anything else on which he feels inclined to digress. This is Burton's "world," and he knows it and loves it from within the security and comfort of his cramped and book-laden cell.

As for me, my browser history reports that last night, between one and two in the morning, I passed, with a few intermediate clicks, from the Wikipedia page on the Kuiper Belt (a part of our solar system beyond Pluto), to crop milk (a secretion regurgitated by parent birds of some species for the feeding of their young), to the Hesychast controversy (a theological dispute in fourteenth-century Byzantium). This was a fairly typical evening for me, part of a circadian rhythm in which my diurnal work, intense and focused, gives way in the night to desultory and stochastic exploration of what I take to be my "world," which includes all the objects in orbit of our sun, all living creatures on earth, all the strange controversies the theologians have taken so seriously, and so much more besides. This tour often tapers off imperceptibly into sleep, and the rhythm begins again the following day, which under this new regime of confinement will almost certainly be nearly identical to the day that preceded it. I am already fearing and mourning my eventual release back into the "world" in the vulgar sense: the bare physical reality of commutes, meetings, dinner parties, activities.

If the most frequently trafficked websites in the world were to be made to stand trial for the damages they have wrought in human history, as we imagined briefly in the introduction, it seems likely that while the harshest charges could be made to stick to Facebook, Amazon, Google, Twitter, and the other usual culprits, Wikipedia might have some hope of being swiftly cleared and released. It is the one large-scale internet project that does not seem to be showing the signs of corruption that have become impossible to deny nearly everywhere else in the past decade, the one part that does not seem to have veered off course from the utopian dream that emerged in early modern Europe of machine-assisted learning for the betterment of humankind.

Over this period of disappointment—roughly the course of the 2010s—in which the arc of an old utopian dream bent toward dystopia, a very different transformation took place in the perception of Wikipedia. Ten years ago, an admission to having consulted this resource would trigger snickering in others, mockery for one's superficial and lazy approach to knowledge. A student who cited Wikipedia would be sternly warned to consult, from now on, only authoritative sources such as books and periodicals. But now the stigma has largely faded. An undergraduate in a typical introductory humanities course is seldom expected at this point to open a book at all, and at more advanced levels of research as well Wikipedia is now mostly seen to be as reliable a starting point as any other reference work for beginning one's research into a topic.

I have already confessed to being addicted to Wikipedia as a form of nocturnal voyaging through the imaginary landscapes of knowledge. More strongly, I will acknowledge that it has also become central to my research. For example, it has furnished me with a list, by year, of the rectors of the University of Halle from its founding in 1502. This is information I found no reason to doubt, and which I therefore used in a book published with Oxford University Press in 2020.[6] And now, in quarantine, it is more vital than ever: some half-remembered point about Turing machines or Aristotelian syllogism can be quickly recalled to mind thanks to this resource, which feels, phenomenologically, much more like a prosthetic memory than like a reference work in the traditional sense. It has fundamentally changed the way I relate to knowledge (or at least to knowledge of particulars). Twenty years ago I could easily have found myself sitting around doing nothing, when the question might suddenly come to me: "What *is* a quasar, anyway?" It is almost certain that at that time I would have quickly abandoned my curiosity, hoping perhaps that I might some day happen upon an answer, but not being quite interested enough, typically, to seek one out. Today it is second nature for me to immediately turn to Jimmy Wales's infinite encyclopedia.

The consequences of this reflexive habit for my general knowledge of the world are profound. I am convinced that

I know vastly more than I would have had this resource not been available to me. In this respect, Wikipedia is the full realization of the dream of the authors of the *Encyclopédie* in the late eighteenth century, who were themselves building upon projects, dating back to the Renaissance, for the schematization and orderly presentation of all of the branches of human knowledge. This project, as pursued for example by Petrus Ramus in the sixteenth century, shared a certain common genetic strand with the various formal-language projects and efforts to develop artificial systems for information-processing that we have already considered in earlier chapters. Wikipedia combines the Renaissance drive for systematization of a Ramus with the project of artificial information-processing as pursued by Leibniz and Pascal, with the massive collaborative approach to learning promoted by thinkers of the Scientific Revolution such as Francis Bacon, and the compendious reader-friendly presentations characteristic of the work of Diderot and D'Alembert on the *Encyclopédie,* rightly seen as the culmination and the defining work of the Enlightenment knowledge project. It is, in short, a delayed achievement of the Enlightenment.

How Wikipedia has managed to synthesize all these strands so successfully, without taking the same troubling turn that social media in general have taken (for Wikipedia's open editing platform makes it a variety of social media) toward disinformation and a general muddying of the distinction between truth and rumor, remains an important question. As Jonathan Zittrain has observed, inverting the old line about Marxism, Wikipedia works well in practice, but not in theory.[7] One of the keys to its success is that the openness of Wikipedia's entries to editing, deletion, and expansion is not a free-for-all. There are at least minimal gatekeeping requirements, as well as the possibility of reversal of new edits by experienced editors, that help to avoid the sort of degeneration, signal loss, digression into irrelevance, and outright vandalism that are so common in comment threads and in public responses to media posts on Facebook and Twitter. The structure of Wikipedia's gatekeeping procedures, combined perhaps with the very nature of the project, sustains something approaching a community spirit, a sincere and non-dogmatic concern to adhere to the truth.

The World Book

Robert Burton was idiosyncratic in numerous ways, but in his desire to apprehend the world from within his cell, to have it delivered up to him in microcosmic form, he shared in the same dream as countless other thinkers across the centuries. Thus, books have commonly been described variously as theaters, as epitomes, as mirrors, and as "counterfeits" (that is, reproductions) of the world, or of the part of the world that they take as their subject. Advances in the technology for printing illustrations in the early modern period deepened metaphors such as these. Few books published in the seventeenth century would enjoy more influence than Jan Amos Comenius's 1658 *Orbis Pictus* [*The World Depicted*], intended as an illustrated textbook for educating children, and translated into dozens of languages over the following two centuries.

In his preface Comenius describes his work as "a little book . . . of no great bulk, yet a brief of the whole world."[8] The very first lesson presents a short dialogue between a master and a boy. The one invites the other to "learn to be wise," which he explains is "to understand rightly, to do rightly, and to speak out rightly all that are necessary."[9] The master promises that with God's help he "will shew thee all."[10] The instruction will begin with "the plain sounds, of which man's speech consisteth," and after this is learned, he promises, "we will go into the World, and we will view all things."[11] The world, here, is nothing other than the pages of the book, with what inevitably appear to us today to be rather crude engravings. But even a rough image of a plough or a ship is quite enough to stimulate the imagination to fill in the missing details and to send the boy outside of himself. That a book is a journey is a cliché so familiar that most of us learn to stop using it by the time we move beyond adolescence. But the disavowal is not a rejection of its essential (if metaphorical) truth.

The worldhood of a book—and here we must resort to this unattractive term of philosophical art, since "worldliness" and "mundanity" have taken on other meanings—and thus the possibility of voyage within it, is what places it in continuity with the oral traditions that long preceded it and with the internet that came after it: all are technologies that aid the human mind in exercising its innate, evolved capacity to get outside of itself (another term for which is "ecstasy"). And this worldhood is all the more evident when it takes the world itself (the world in the sense of the totality of navigable regions of the surface of the earth) as its object, and, as its goal, the scale representation in microcosmic form of that totality. Renaissance and early modern mapmakers frequently thematized this microcosmic character of cartography's productions, and these were often given names that suggested they were more than just books, but rather "theaters" of the world, such as Abraham Ortelius's influential *Theatrum Orbis Terrarum* of 1570. In their perceived nature and in their function, these books, as much as they have in common with other books that invite their reader on a sort of voyage, exist in even closer continuity with sundry non-book objects, such as globes (both terrestrial and celestial), astrolabes, and all the various "living instruments" of early modern science, some of which existed and some of which were merely mythical, that combined the principles of mechanics and alchemy in order to somehow capture in an enclosed space some principle, force, or structure that pervades or characterizes the world beyond the laboratory.

In all cases, the aspiration was to capture at the human scale what ordinarily exists only at the cosmic or at least the global scale. In the pursuit of this aspiration, the boundary between representation and containment was often obscure. That is, whether an element of the outside world that is present in the technological instrument is present literally, as a trace or sample, or whether it is only present symbolically, as an image, is not always made clear by the inventor. Sometimes, we find, an invention amounts to a curious hybridization of both "real" and representational components. Thus, in 1711, Leibniz writes a letter to Duke Anton Ulrich of Braunschweig-Wolfenbüttel in which he describes a plan to impress Tsar Peter the Great of Russia on an upcoming diplomatic visit to Germany. The philosopher proposes that while dining, instead of the usual dinner theater [*Comoedien*], the hosts might instead arrange "a presentation of the power of the Great Tsar through a representation of his Empire, with carved figures that variously serve this purpose and indicate the Tsar's military victories."[12]

So far, what Leibniz imagines is only a large relief map, and thus a pure representation. But he adds that it would ideally be something more than that too. He recommends positioning it on the floor of a ballroom adjacent to the dining table, so that water could easily be carried to it. And then he explains what the water is for. "So that it should contain within it a new and special contrivance," he writes, "the representation of the great Muscovite Empire should not be a mere surface, as in land maps, but rather a true depiction of the heights and depths of the land, the source of the streams, showing their course and flow into the sea, to which end it should be ensured that the five seas are at the lowest place, and the highlands at the highest, and everything in between."[13] Thus, in Leibniz's vision, the miniature Dvina River, the Ob, the Yenisei, and the Lena "should all flow into the White Sea, the Narva and the Duna should flow into the Baltic, the Volga into the Black Sea, and the Amur into the sea of Japan."[14]

All this may seem fanciful, even by the standards of early modern court culture, but whatever the plan's extravagance, it also shows Leibniz the courtier's clear understanding of the intimate relationship between verisimilitude in representation, on the one hand, and political power on the other. Even where there is no flowing water, maps of a sovereign's territory have often been seen not just as microcosms, but as condensed epitomes of his reign. No area of a palace is better designed to reassure a sovereign of his hold on power than the map room. Even if until the past century or so such a room could not give a leader any immediate power of action (Peter the Great could not, say, order a drone strike on an unruly corner of his empire while studying a representation of it on the map-room wall), nonetheless there is nothing quite like visualization—of a space, in the case we are discussing here, but certainly also of data—to make a person feel on top of things. Leibniz's added twist, in the case of the relief map he would have had built in Wolfenbüttel (the plan never materialized), is to meld the representational and the real, to seize onto the real principles of hydraulics to cause the water to flow on the map of Russia just as it does in Russia itself.

But why go to such trouble? In the study of aesthetics philosophers sometimes seek to define art narrowly so as to exclude direct encounter with elements of physical reality, in contrast with transformations of physical reality that point to something else (recall the earlier discussion of "aboutness"). Thus we sometimes suppose that painting and sculpture are art in the highest sense, because they are at least traditionally not "about" oil or marble (all this changes in the mid-twentieth century, with iconic works such as Jasper Johns's tellingly named *Canvas* of 1956). Rather, oil and marble are used in the course of creating something that is about, say, St. Sebastian or Aphrodite. Perfumery, by contrast, is not about anything other than the molecules of perfume getting into your nose and creating a direct sensation of them.

Of course, the line is blurry, and one could argue that odor's power to induce memories is denotative in the same way other art forms are (not to mention instrumental music, which often seems to be about something, even if it is difficult to say what that something is). The distinction becomes rather difficult to uphold when we consider the very many examples of traditional art and craft that share in the essence or nature of the thing they represent, either in fact or in the way they are conceived in the culture that produces them: for example, Melanesian ritual masks or effigies that are made with human hair or fingernails, or consecrated icons that not only depict sacred things, but that are held to be sacred in themselves. It should not be so surprising to find that the line is similarly blurry when we move out from "art" in the narrow sense to "artifice" in the broad sense, including, notably, the scientific instruments and models of early modern Europe.

If Leibniz's hydrodynamic relief map is not enough to establish the continuity of early modern scientific devices with magical craft artifacts such as so-called voodoo dolls, we might also consider the example of Cornelis Drebbel's "weather globes." A seventeenth-century Dutch alchemist who succeeded in designing a functional submarine, and claimed to have succeeded in transmuting base metal into gold, as Vera Keller has shown, Drebbel also sought to develop what she terms a "cosmoscope": "a single living machine that could encapsulate, prove, and effortlessly convey universal knowledge of nature."[15] Less discerning historians have imagined that Drebbel's device was simply something like an early attempt at a thermometer, and indeed it was at least that. But it was also meant by its inventor to do something vastly more ambitious than to simply report the ambient temperature: it was to harness the same cosmic principles that govern all of nature, and thereby to instantiate the perpetual motion of the world itself.

One of the functions of such a device would be to reproduce weather phenomena—not simply to report the temperature, but to literally encapsulate within an observable glass container the same meteorological processes that make the external air hot or cold, humid or dry. According to some even more fanciful variations on the idea of the weather-glass or *vitrum calendare* (a term that is frequently used by Francis Bacon), the glass-enclosed chamber of the device would be something like the snow-globe souvenirs we know today, showing all manner of meteorological phenomena, including the formation of clouds, snow, rain, lightning, and so on, all on the natural philosopher's desktop.

Do We See through the Internet?

This brief foray into the history of early modern scientific instruments might help us better to understand what we have been calling the "microcosmic" character of the internet. We ordinarily think of the internet as transmitting only representations, indeed symbolic representations that require interpretation at the one or the other end of the transmission. This is the way I have been characterizing the internet throughout this book as well. Electrical pulses arrive at a terminal point of the internet and on the screen at that point they generate an image of a flamingo, but in their transit they do not resemble a flamingo at all, let alone do they share in the nature of an actual flamingo.

For the most part, all we have with the internet are such representations on screens, in written and spoken words, and in still and moving images. Yet there are also respects in which we may be said to be experiencing "the real thing" when we are on the internet. 3D printing is increasingly making it possible to pull, say, an actual gun out of the internet, rather than simply learning about guns or watching representations of what they do. The burgeoning field of "teledildonics" is making it possible to bring about not just visual and aural transformations at a distance for a conversation partner, but also haptic or vibratory manipulations of a sexual partner.

But beyond these outlying cases, there are other, much more common ones. On most understandings of the ontology of a musical work, when we hear music streaming through Spotify or YouTube, we are hearing the music itself, and not just a representation of it. And much less or much more mysteriously (depending on your philosophical commitments), when we pay bills online, when we issue such speech acts as promises or threats via Facebook or Twitter, when we tell our loved ones we love them over Skype or Zoom, we are bringing about real transformations in the world, in our financial situation, in our social standing, in our hearts. All of this is conducted through electrical pulses, but it is not all mere "simulation." The remote-controlled vibrator or the 3D-printed gun are only further twists on a power we had already mastered and were already exercising, to effect world-changing action at a distance.

Nor is it clear that the manner in which internet-mediated representations are delivered to us as users distinguishes this tool in any significant way from other scientific instruments that are sometimes thought to give us more immediate access to the thing we wish to observe. It is a commonplace among philosophers of science that no easy line can be drawn between direct and indirect observation of an external thing. Even unenhanced vision is mediated by light rays, and common telescopes, while they enable us to see things we would not ordinarily see, rely on optical principles that are substantially the same as those of the anatomy of the eye.

Yet for a long time now telescopes and microscopes have not simply been enabling us to "see through" them as we would through our bifocals. Rather, they are complex machines, often relying on principles other than or in addition to the physics of light and the physiology of vision, in order to generate an interpretable representation of the invisible object of study. When an electron microscope yields a representation of the microstructure of molecules, or a telescope reports back about distant galaxy clusters, the "report" in question is one that looks no different to the scientist interpreting it on a computer screen in a laboratory or an observatory, than it does to a person sitting at home seeing the same report transmitted over the internet. For most purposes, an astronomer could perfectly well "work from home," and in so doing could "see" all the same things through the internet that she would see if she went to the observatory.

To revive the philosopher of science Ian Hacking's famous question in his article, "Do We See through a Microscope?" we may also ask: "Do we see through the internet?" And the most appropriate answer seems to be: "At least as much as we do through a microscope."[16] That is, the way the microscope—like most other complex scientific instruments today—yields up information about the world around us is no more direct or unmediated a way than the internet yields up such information. Again, in many cases the report from such an instrument that the scientist is reading from a screen in a laboratory is the same as the report that I could also view on my screen at home, or that the scientist could have received on her screen, with all the right top-secret passwords, if she had stayed home. If I want access to the Kuiper Belt, if I want to do the closest thing to voyaging to the Kuiper Belt of which humanity is at present capable, I am for most purposes just as well positioned to set out on such a voyage at home as I am at the world's most advanced observatories.

It might seem that this ability, to voyage while remaining at home, is one that has only arrived with the most recent technologies, but in a well-appointed cell such as Burton's, with multiple volumes of works of cosmography, the experience was no doubt not so unlike my own late-night visits to websites with NASA satellite images. Just as we know that the neurological activity associated with immersive cinema or virtual reality is not very different from what goes on in the brain of, say, a Paleolithic child listening to a story told in the depths of a cave before a flickering fire, so too in the case of books and the internet does the common substrate of imagination ensure that the experience is substantially the same no matter what the level of the technology that is aiding us in exercising the imagination. Here we have occasion to correct a basic fallacy in the way we often think about technological progress. It is perhaps not that our new inventions *provide* us a new sort of experience, but rather that we come up with ever new ways of activating the sort of experience we as human beings have always had. No one would have thought to put the physics of light to use for cinematic storytelling, if we did not already know intimately the sort of experience that such storytelling provides. And we know it intimately because of our experience of storytelling in its purest and most elementary sense.

But as to the similarity between Burton's perusal of his cosmographies and mine of the NASA website and all its rich public-directed resources, the fact that our access to outer space is and always has been a sort of imaginative journey activated by whatever aids to the imagination are available to hand, is perhaps easier to see than when we read about, say, the physiology of birds or the architecture of cathedrals. There is a sense in reading about these latter things that if we come to be deeply interested in them, we must put down the books at some point and go have a look at the real thing, which, unlike the Kuiper Belt objects and the stars beyond them, we are in fact capable of studying up close. And yet the book is not simply a consolation for the inaccessibility of the thing that interests us. Sometimes it is a supplement, giving us a different sort of access to the thing that we might also study up close. It is never a replacement for the thing, though it may at least be a temporary stand-in for the thing until we figure out how to access it (either for reasons of physical distance, such as the Kuiper Belt, or because its nature is not known and its very existence is in doubt, such as God, numbers, justice, and all the other usual objects of philosophy). And while it is never a permanent replacement, a book is often not just a supplement, but indeed a sublimation of the thing it is about, or is often experienced as such: as the channel by which a mundane, common, or grotesque feature of reality is made wonderful. I for one certainly do not wish to see a bird up close regurgitating its crop milk, but to read about this is to induce a feeling of awe at the variety and intricacy of nature's workings.

Nor is it simply that the mediation of the text provides me with a safe distance, a buffer, that keeps the bare reality of some potentially disgusting or overwhelming thing from getting too close. It is rather that engaging with the thing through this mediation also makes it possible to place it in relation to other things, things that would not be present "in the field." On the internet this placing-in-relation is often facilitated by hyperlinks. On the Wikipedia page about avian crop milk, I click the link to the page about the *Symphysodon* genus, also known as "discus fish," some species of which "nurse" their young on a secretion through the skin that has molecular properties similiar to milk. And as I read, I begin to wonder: just what is milk anyhow? And how widely is it distributed throughout living nature? If I am not cautious I will soon find myself reading about the Milky Way and the metaphors that first associated this star system with milk, but I restrain myself and I think about animal secretions only, and even here, under these constraints, I am aware of a rising sensation in me of the sublime.

The Infinite Book Wheel

In his 1588 work, *The Diverse and Ingenious Machines of Captain Agostino Ramelli*, the eponymous Italian military engineer described, alongside 194 other designs, what he called the "book wheel." Working in a tradition of speculative invention that extends back to Leonardo da Vinci and that often involves the imaginative description and illustration of devices that may or may not be realizable in the physical world (think of Leonardo's helicopter, which was to have become airborne from the constant turning of cranks by four men standing on a heavy wooden platform), Ramelli proposed numerous original devices driven by the force of springs, levers, hydraulics, and other mechanical principles. A good number of his designs, as is typical of the tradition in which he is working, are for machines of military

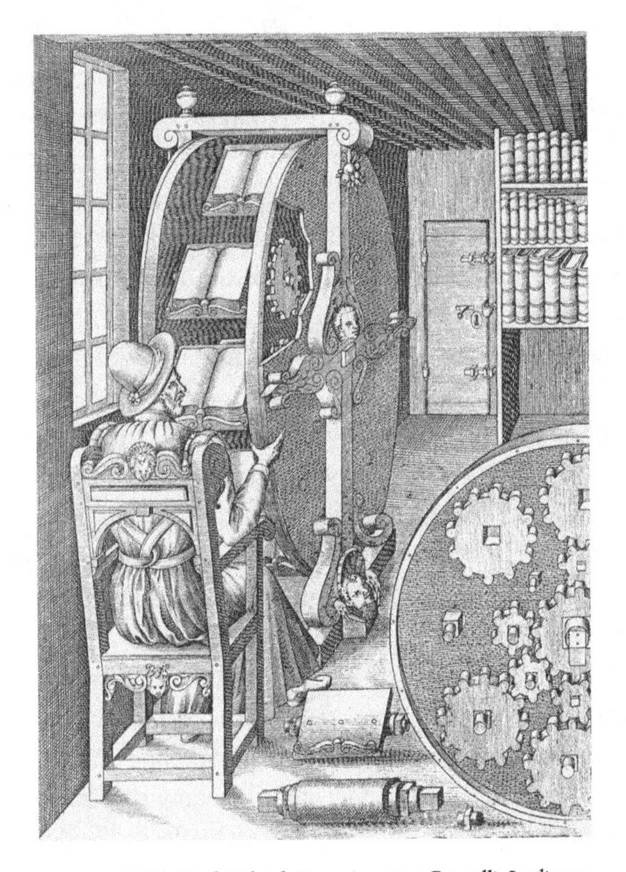

FIGURE 6. The Book Wheel. From Agostino Ramelli, *Le diverse et artificiose machine*, 1588.

aggression: new sorts of catapults, battering rams, and other siege engines. Just as the internet has developed in conjunction with defense initiatives, so too does the author of these diverse and ingenious machines place his information-processing device within a broader landscape of weapons of war.

Not everyone is fit for war, however, and Ramelli imagines that the book wheel is ideally suited for "any person who delights in study, principally those who are indisposed or subject to gout."[17] With such a machine, he writes, "a man can look at and read a great quantity of books without moving from one place."[18] The machine is constructed in such a way that "when books are placed on the tablets, however much one turns said wheel around, said books will never fall or slip from the place in which they have been put, thus remaining always in the same state, and presenting themselves to the reader in the same way."[19] In the following century the French inventor Nicolas Grollier de Servière appears to have built a working model of Ramelli's device, described in a posthumous catalogue of his inventions as a "Particular sort of reading desk, very suitable for learned people" (Pupitre d'une façon particulière, & très commode pour les gens d'étude).[20]

Many commentators have noticed the hypertext-like possibilities of such a device. We might imagine loading it with one particularly difficult text whose concepts and vocabulary have not all been mastered by the reader, and then also with several other reference works, open to pages that explicate these concepts and vocabulary. The reader will then, upon arriving at a difficult point in the main text, simply manipulate a lever that will mechanically pull up another work to clarify the difficulty. The experience would thus, we can imagine, be very much like the experience of reading about crop milk on Wikipedia, arriving at a hyperlinked reference to something called "prolactin," pausing to click on it in order to learn that it is a protein responsible for triggering the production of milk, and then returning to the original article.

Of course the book wheel is self-contained and obviously very limited, while the internet is practically infinite in the possible hyperlinked pathways it presents to us. Yet a dozen or so books on a wheel are already enough to stimulate the idea of a possible extension of the device's powers ad infinitum, and from the Renaissance onward this invention would continue to develop as one of the genetic strands that eventually merged with the telecommunication and computing technologies that were evolving in parallel, in order to become the internet.

There is a certain irony in the fact that part of the long history of the internet involves efforts to accommodate gouty readers. An affliction that at least seemed much more widespread in the early modern period than today, gout was associated with the aristocracy, but also with everyone, including scholars, who leads a physically inactive and sedentary life. Today the notorious conditions associated with excessive internet use—diabetes, obesity, high blood pressure—are different, but the general problem has not yet been solved: that exploration of virtual microcosms of the world does little to preserve our bodies, condemned as they are ever to exist in the macrocosmic physical world.

Learning, an old cliché had it, is an *ambulatio animae*, a stroll taken by the soul. But this is a stroll tracked by no FitBit or other step counter, and the way it neglects the body is both a standard trope and stereotype about philosophers and scholars (with many philosophers straightforwardly denying that the well-being of their bodies had anything to do with their own well-being: "Pound, pound the pouch containing Anaxarchus," said Anaxarchus, "you do not pound Anaxarchus"[21]), and also a serious public health concern of the present moment. Disembodied experience—as envisioned in folk culture, for example through the figure of the *benandanti* or night-walkers, who go out voyaging over great distances, as their bodies lie asleep at home—is a dream that is at least partially realized in any experience of imaginative reverie, and the realization seems all the more veracious when stimulated by a book or by a screen attached to the internet.

For the most part, however, in the more staid settings of academic philosophy and theology, the accepted view is that, other than perhaps in death, the soul must stay in the body. And for philosophers this is so whether the soul is conceived as a distinct substance that shares nothing of the nature of the body (as in Descartes), or as a subtle or ethereal principle that permeates the body (as in the Stoic philosophers or the Cambridge Platonists), or as an emergent consequence of the activity of the body, which is in itself strictly physical. No matter what the particular ontology, for the most part philosophers have all been in agreement that any *ambulatio animae*, any wandering of the soul, is only metaphorical.

And yet the soul might well seem to migrate *within* the body, depending on the sort of bodily activity a person undertakes or seeks to undertake (a remnant of this idea echoes in the common observation that a promiscuous man is "thinking with" his sexual organ), and a thinker who is interested in a particular kind of human activity will be correspondingly more inclined to anchor the soul to the region of the body associated with that activity.

In *The Anatomy of Melancholy*, Burton incorrectly attributes to the first-century BCE Roman historian Sallust a curious statement: "They have the seat of their soul in the tips of their fingers."[22] In Burton's own day, the search for the *sedes animae* was an important one for medicine, philosophy, and that large grey area that still existed between these two fields. Aristotle had held that the seat of the soul was the heart; the Jewish esoteric tradition, of which Burton was aware, sometimes held that it was a bone at the base of the spinal column; some thinkers in the period were coming around to the view that it was in the brain; soon after Burton wrote, René Descartes would propose, drawing on long physiological tradition, that the seat of the soul is in the pineal gland. Burton appears to believe that Sallust had made his comment in reference to artisans "expert in manufactures," and implies that the hand's ability to work as if on its own is at least as good a ground for locating the soul or mind in these extremities than, as is usually done, somewhere closer to the center, as in the heart or the brain, closer to the center of language production in the human body.

Although others, of a younger generation or of a different social background, often manage to communicate through the internet while largely circumventing keyboards, for me and many others like me, the experience of being online is mediated almost by definition through typing, pressing down an arrangement of keys representing the elements of our alphabetic language, and originally set in their unusual analphabetic order, the beloved QWERTY, in 1878, in order to prevent the bars of the typewriter from jamming.

Today there is no such danger, and when I type, the motion, the flow from my thoughts to the tips of my fingers through the keyboard and into the document on the screen, is as fluid and easy as if I were speaking. As I am typing at this very moment, I am not looking at the computer at all, but out the window at the leaves on the trees. That is how intimately I know my machine, and how natural it is for me to think and to produce traces of my thoughts by means of it. If we must find a seat for the soul, surely the tips of the fingers are as attractive a location for this noble role as any other.

When we were children with our Atari game modules, there was a common facetious argument produced in opposition to our parents' orders to stop playing video games and to go outside: that these games were beneficial for the development of "hand-eye coordination." Now many decades later I find there is a perfect coordination, almost like ballet, in my own human-computer interface, of not just hand and eye, but of hand, eye, and world.

I type the phrase "Kuiper Belt" as quickly as I can think it, as quickly as I can perceive the desire to absorb the facts of it, and no less quickly do the facts come pouring in from my screen. It is a dream come true, this cosmic window I am perched up against, this microscosmic sliver of all things.

NOTES

Introduction. "Let us calculate!"

1. Leibniz, *Die philosophischen Schriften von G. W. Leibniz*, vol. 7, 125.

2. See Soto, *Esperanto and Its Rivals*.

3. Wiener, *Cybernetics*, 175.

4. Ibid.

5. Stanley, *The Technological Conscience*, xi.

6. Ibid., xiii.

7. Ibid.

8. Walzer, "Town Meetings and Workers' Control," 273–90, 274.

9. Hacking, *Historical Ontology*.

Chapter 1. A Sudden Acceleration

1. Tweet from @getfiscal, dated December 23, 2020. https://twitter.com/getfiscal/status/1341730505528659968

2. Serres, "Internet, c'est du Leibniz sans Dieu."

3. Citton, *The Ecology of Attention*. For further studies of the problem of attention in its cultural and social dimensions, see Crary, *Suspensions of Perception*; Wu, *The Attention Merchants*. For a masterful treatment of the philosophical problems of the faculty of attention, see Montemayor and Haladjian, *Consciousness, Attention*. For a historical treatment of the topic of attention in its relation to the natural philosophy of the early modern period, which has frequently been of interest to us throughout this book, see Daston, "Attention and the Values of Nature."

4. "The Porn Machine," *n +1 Magazine* 5 (Winter 2007).

5. Nabokov, *Pnin*.

6. See James, *Principles of Psychology*, chapter 11.

7. Dicey Jennings, *The Attending Mind*.

8. Ibid., 3.

9. Ibid., 2.

10. Ibid., 8.

11. See Kim and Cave, "Perceptual Grouping"; cited in Dicey Jennings, *The Attending Mind*, 8.

12. Ganeri, *Attention, Not Self.*

13. Ibid., 1.

14. Ibid., 12.

15. Ibid.

16. Ibid., 3.

17. Ibid., 4.

18. See Lotze, *Geschichte der Aesthetik in Deutschland.*

19. See Robert Vischer, *Über das optische Formgefühl. Ein Beitrag zur Ästhetik.*

20. See Friedrich Theodor Vischer, *Aesthetik oder Wissenschaft des Schönen.*

21. See in particular Latour, "Why Has Critique Run Out of Steam?"

22. See in particular Heidegger, "The Question Concerning Technology."

23. See Malcolm Foster, "Aging Japan: Robots May Have Role in Future of Elder Care," *Reuters*, March 28, 2018. https://www.reuters.com/article/us-japan-ageing-robots-widerimage-idUSKBN1H33AB

24. Tweet from @balajis, dated April 29, 2021. https://twitter.com/balajis/status/1387620880893681670

25. Williams, *Stand Out of Our Light*, 12.

26. See Blair, *Too Much to Know.*

27. See Yates, *The Art of Memory.*

28. For an important study of the role of attention in classical Indian philosophy, see Ganeri, *Attention, Not Self.*

29. This language has also been retained and promoted by at least one group of art appreciators. See in particular the discussion of "birds" at the thirty-third annual São Paulo Biennial, as related by the biennial's curator, Gabriel Pérez-Barreiro: Seph Rodney, "At This Year's São Paulo Biennial the Curator Prioritizes Feeling Over Discourse,"

30. Williams, *Stand Out of Our Light*, 12.

31. Leopold, *A Sand County Almanac*, 215–216.

32. Lanier, *You Are Not a Gadget.*

33. See "Our Statement," at https://www.somanyofus.com.

34. See Charlie Warzel, "Is QAnon the Most Dangerous Conspiracy Theory of the 21st Century?" *New York Times*, August 4, 2020. https://www.nytimes.com/2020/08/04/opinion/qanon-conspiracy-theory-arg.html

35. Lynn, "The Big Tech Extortion Racket."

36. For an engaging analysis of this phenomenon, see Vanderbilt, *You May Also Like.*

37. Niall Ferguson, "TikTok Is Inane. China's Imperial Ambition Is Not," *Bloomberg Opinion*, August 9, 2020. https://www.bloomberg.com/opinion/articles/2020-08-09/tiktok-is-the-superweapon-in-china-s-cultural-warfare?sref=ojq9DljU

38. Lanier, *Ten Arguments*.

39. Tweet from the @AliceFromQueens account, dated May 13, 2019.

Chapter 2. The Ecology of the Internet

1. Farag, Zhang, and Ryu, "Dynamic Chemical Communication."

2. For a classic treatment of technology as an "extension," though one that does not dwell on parallels to phenomena of growth in the natural world (and that generally addresses questions in the philosophy of technology from a human-exceptionalist point of view), see McLuhan, *Understanding Media*.

3. Marcus Aurelius Antoninus, *Meditations*, 91.

4. Lucian, *A True History*, 35.

5. See *Curious Enquiries*.

6. Ibid., 2.

7. See Justin E. H. Smith, "The Internet of Snails."

8. See Japyassú and Laland, "Extended Spider Cognition." For the classic treatment of "active externalism" in human cognition, which supposes that, like spiders, our mental activity similarly reaches out into the world around us, particularly into our technological prostheses, see Clark and Chalmers, "The Extended Mind." See also the collection of papers, including a reprint of Clark and Chalmers's article, edited by Menary, *The Extended Mind*.

9. See Whiteside et al., "Mycorhizal Fungi Respond to Resource Inequality."

10. See Buchanan, "Deleuze and the Internet," 1–19, 9–10.

11. See in particular Mancuso, *The Revolutionary Genius of Plants*.

12. McFarlane, "The Secrets of the Wood Wide Web."

13. See Kant, *Critique of the Power of Judgment*, §61, 5:360/264.

14. For a critical discussion of this contention, see Piercey-Normore and Deduke, "Fungal Farmer or Algal Escorts."

15. La Rue, *Report of the Special Rapporteur*.

16. For an excellent philosophical genealogy of communications theory and its practical applications in the nineteenth century, see Kittler, *Discourse Networks 1800/1900*.

17. See Mesmer, *Mémoire sur la découverte du magnétisme animal*.

18. See Digby, *Discours fait en une célèbre assemblée*.

19. Heinrich Heine, cited in Figes, *The Europeans*.

20. Gruzinski, *Quelle heure est-il là-bas?*.

21. Colin Renfrew, "Neuroscience, Evolution, and the Sapient Paradox: The Factuality of Value and of the Sacred," *Philosophical Transactions B* 363, /499 (June, 2008): 2041–2047, 2041.

22. "Les Nouvelles admirables," in L'Hermite (ed.), *Recueil de pieces en prose*, 381; see also Sutton and Sutton, "The Recording Sponge." The authors date the original text, on which the one in L'Hermite's volume is based, to the early 1630s.

23. White, "Energy and the Evolution of Culture," 338.

24. Shapin, *The Scientific Revolution*, 1.

25. Scott, *Against the Grain.*

26. Balée, *Cultural Forests of the Amazon.*

27. Kant, *Critique of the Power of Judgment* §64, 5:370/242.

Chapter 3. The Reckoning Engine and the Thinking Machine

1. Alim et al., "Mechanism of Signal Propagation in *Physarum polycephalum.*" See also Tero et al., "Rules for Biologically Inspired Adaptive Network Design."

2. Bostrom, *Superintelligence*, 164.

3. See Corey S. Powell, "Elon Musk Says We May Live in a Simulation. Here's How We Might Tell If He's Right," *NBC News*, October 2, 2018.

4. Kant, *Critique of the Power of Judgment*, 5:464/328.

5. Bostrom, "Are You Living in a Computer Simulation?", 243.

6. Dennett, "Will AI Achieve Consciousness? Wrong Question."

7. Schneider, "It May Not Feel Like Anything to Be an Alien."

8. See Kurzweil, *The Singularity Is Near*, 136. "I set the date for the Singularity—representing a profound and disruptive transformation in human capability—as 2045."

9. Schneider, "It May Not Feel Like Anything to Be an Alien."

10. For an extensive discussion of Rorario's treatise, see Justin E. H. Smith, *Irrationality.*

11. Brian Cantwell Smith, *The Promise of Artificial Intelligence*, xv.

12. See Adams and Browning (eds.), *Giving a Damn.*

13. Cantwell Smith, *The Promise of Artificial Intelligence*, 108.

14. See for example Carruthers and Smith, *Theories of Theories of Mind*; Winfield, "Experiments in Artificial Theory of Mind"; Chatila et al., "Toward Self-Aware Robots."

15. Prashanth Ramakrishna, "'There's Just No Doubt That It Will Change the World': David Chalmers on V.R. and A.I.," *New York Times*, June 18, 2019.

16. John Basl and Eric Schwitzgebel, "AIs Should Have the Same Protections as Animals," *Aeon*, April 26, 2019.

17. Namely, Eric Schwitzgebel, as related in private correspondence.

18. Bacon, *The Cure of Old Age, and Preservation of Youth*, Preface, n.p.

19. Ibid.

20. Ibid.

21. Chistiakov, "Narodnoe predanie o Briuse (iz vospominaniï moego tovarishcha)."

22. Leibniz, "Machina arithmetica," 181.

23. See in particular Pérez-Ramos, *Francis Bacon's Idea of Science.*

24. Leibniz, *Monadology*, in *Die philosophischen Schriften von G. W. Leibniz*, vol. 6, 609.

25. Ibid.

26. Davis's version of the thought experiment was presented in a seminar at MIT in 1974, and reported only later by others in attendance. See in particular Dennett, "Toward a Cognitive Theory of Consciousness."

27. Block, "Troubles with Functionalism."

28. See Searle, "Minds, Brains, and Programs."

29. Leibniz, *Novissima Sinica Historiam nostri temporis illustratura*, in *Gottfried Wilhelm Leibniz: Sämtliche Schriften und Briefe*, Series IV, vol. 5, 384.

30. Fage, *A Description of the Whole World*, 49.

31. See S. Korsakof [Semyon Korsakov], *Aperçu d'un procédé nouveau d'investigation.*

32. Ibid.

33. Ibid.

34. Ibid.

35. Ibid.

36. For a fine survey of the theoretical aims of artificial-language schemes in England in the seventeenth century, see Lewis, *Language, Mind and Nature.*

37. Leibniz, "Machina arithmetica in qua non additio tantum et subtractio . . ." 180.

38. Wiener, *Cybernetics*, 41.

39. Ibid.

40. Ibid.

41. Ibid.

42. Ibid.

43. Ibid.

44. Ibid.

45. Ibid.

Chapter 4. "How closely woven the web": The Internet as Loom

1. More, *The Immortality of the Soul*, Book 3, 280.

2. See https://twitter.com/iam_johnw/status/1268754495997165568.

3. Marcus Aurelius Antoninus, *Meditations*, 91.

4. Müller (tr. and ed.), *The Upanishads*, Part 2: *Brihadāranyaka Upanishad*, Adhyāya 3, Brāhmana 8, 137.

5. Much excellent scholarship has been produced on the connection between the history of textile technologies and the history of applied mathematics broadly speaking. Here I am especially indebted to Bredekamp, "Leibniz' Gewebe: Strumpfband, Falte, Leinwand"; Brezine, "Algorithms and Automation"; Friedman, "Baroqueian Folds"; Harlizius-Klück, *Weberei als episteme*. Among these authors, Friedman is engaged in ongoing research on the intersection of weaving and mathematics in the seventeenth century, not only in Leibniz but as in Joachim Jungius and J. J. Becher. The results of this research are eagerly awaited.

6. Joseph Marie Jacquard, *Métier à tisser*, inventory number 07641–0001, Permanent collection, Musée des Arts et Métiers, Paris. For a comprehensive study of Jacquard's invention in its relation to the history of computer science, see James Essinger, *Jacquard's Web*. See also Davis and Davis, "Mistaken Ancestry"; Langlois, "Distributed Intelligence."

7. See Riskin, *The Restless Clock*.

8. See Li, ". . . 'früh bestelle, und etwas hinein bringe, so geld bringe!' Leibniz' Garten in Hannover," 25–58.

9. Menabrea, "Sketch of the Analytical Engine," vol. 3, 669. Originally published in *Bibliothèque Universelle de Genève* 82 (October 1842).

10. Ibid., 677.

11. Ibid.

12. Ibid.

13. Ibid.

14. Ibid., 678.

15. Ibid.

16. Ibid.

17. Ibid.

18. Lovelace, "Notes by the Translator," 696.

19. Ibid.

20. Ibid. For a discussion of Lovelace's account of the Analytical Engine that places it in relation to Jacquard's loom, among other earlier mechanisms of doing and making, see Schaffer, "Babbage's Dancer," 53–80.

21. Ibid.

22. Ibid.

23. G. W. Leibniz to Joachim Bouvet, February 15, 1701, No. 202, in Leibniz, *Sämtliche Schriften und Briefe*, ed. Berlin-Brandenburgische Akademie der Wissenschaften und die Akademie der Wissenschaften zu Göttingen, Series 1, Vol. 19, Berlin, 2005, 404–405.

24. Kant, *Critique of the Power of Judgment* §90, 5:461/325.

25. See Linus, *Tractatus de corporum inseparabilitate.*

26. Le Grand, *Institutio philosophiae secundum*, 350.

27. Dennis Overbye, "Beyond the Milky Way, a Galactic Wall," *New York Times*, July 10, 2020.

28. Ricoeur, *The Rule of Metaphor*, 300.

Chapter 5. A Window on the World

1. Auerbach, *Mimesis.*

2. Gass, *A Temple of Texts*, 122–123.

3. Burton, *The Anatomy of Melancholy*, 29.

4. Ibid.

5. Ibid.

6. See Menn and Smith, *Anton Wilhelm Amo's Philosophical Dissertations on Mind and Body.*

7. Jonathan Zittrain, tweet dated August, 27, 2018. https://twitter.com/zittrain/status/1033897474849689606

8. Jan Amos Comenius, *Orbis Pictus*, xiv.

9. Ibid., 1.

10. Ibid.

11. Ibid.

12. G. W. Leibniz, Letter to Duke Anton Ulrich, September 1, 1711, in V. I. Ger'e (ed.), *Sbornik pisem i memorialov Leibnitsa otnosyashchikhsya k Rossii i Petru Velikomu*, Saint Petersburg, 1873, 169–170.

13. Ibid.

14. Ibid.

15. Keller, "Drebbel's Living Instruments," 41.

16. See Hacking, "Do We See through a Microscope?"

17. Ramelli, *Le diverse et artificiose machine del Capitano Agostino Ramelli*, ch. 188, 316.

18. Ibid.

19. Ibid.

20. See Grollier de Servière, *Recueil d'ouvrages curieux de mathématiques et de mécanique.*

21. See Laertius, *Lives of Eminent Philosophers*, ch. 9.10, 470.

22. Burton, *The Anatomy of Melancholy*, 113.

GENERAL BIBLIOGRAPHY

Zed Adams and Jacob Browning (eds.), *Giving a Damn: Essays in Dialogue with John Haugeland*, Cambridge, MA: MIT Press, 2016.

Karen Alim, Natalie Andrew, Anne Pringle, and Michael P. Brenner, "Mechanism of Signal Propagation in *Physarum polycephalum*," *PNAS* 114, 20 (May 2017): 5136–5141.

Marcus Aurelius Antoninus, *Meditations* 4, 40, ed. and trans. C. R. Haines, Loeb Classical Library 58, Cambridge, MA: Harvard University Press, 1916.

Erich Auerbach, *Mimesis: The Representation of Reality in Western Literature*, trans. Willard Trask, Princeton, NJ: Princeton University Press, 2003 [1946].

Roger Bacon, *The Cure of Old Age, and Preservation of Youth*, translated out of Latin; with Annotations, and an Account of his Life and Writings, by Richard Browne, London: Thomas Flesher, 1683.

William Balée, *Cultural Forests of the Amazon: A Historical Ecology of People and Their Landscapes*, Tuscaloosa: University of Alabama Press, 2013.

John Basl and Eric Schwitzgebel, "AIs Should Have the Same Protections as Animals," *Aeon*, April 26, 2019.

Ann M. Blair, *Too Much to Know: Managing Scholarly Information Before the Information Age*, New Haven, CT: Yale University Press, 2010.

Ned Block, "Troubles with Functionalism," *Minnesota Studies in the Philosophy of Science* 9 (1978): 261–325.

Nick Bostrom, "Are You Living in a Computer Simulation?" *Philosophical Quarterly* 53, 211 (2003): 243–255.

Nick Bostrom, *Superintelligence: Paths, Dangers, Strategies*, New York: Oxford University Press, 2014.

Horst Bredekamp, "Leibniz' Gewebe: Strumpband, Falte, Leinwand," in R. Hoppe-Sailer, C. Volkenandt, and G. Winter (eds.), *Logik der Bilder. Präsenz—Repräsentation—Erkenntnis*, Bonn: Reimer, 2005, 233–238.

Carrie Brezine, "Algorithms and Automation: The Production of Mathematics and Textiles," in Eleanor Robson and Jacqueline Stedall (eds.), *The Oxford Handbook of the History of Mathematics*, Oxford: Oxford University Press, 2009, 468–492.

Ian Buchanan, "Deleuze and the Internet," *Australian Humanities Review* 43 (2007): 1–19.

Robert Burton, *The Anatomy of Melancholy*, New York: W. J. Widdleton, 1875 [1621].

Peter Carruthers and Peter K. Smith, *Theories of Theories of Mind*, Cambridge: Cambridge University Press, 1996.

Raja Chatila, et al., "Toward Self-Aware Robots," *Frontiers in Robotics and AI*, August 13, 2018.

M. B. Chistiakov, "Narodnoe predanie o Briuse (iz vospominaniï moego tovarishcha)," *Russkaia starina* 8, 4 (1871): 167–170.

Yves Citton, *The Ecology of Attention*, trans. Barnaby Norman, Cambridge and Malden: Polity Press, 2017 [2014].

Andy Clark and David Chalmers, "The Extended Mind," *Analysis* 58, 1 (January 1998): 7–19.

Jan Amos Comenius, *Orbis Pictus*, Syracuse, NY: C. W. Bardeen, 1887 [1658].

Jonathan Crary, *Suspensions of Perception: Attention, Spectacle, and Modern Culture*, Cambridge, MA: MIT Press, 1999.

Curious Enquiries: Being Six Brief Discourses, viz. I. Of the Longitude. II. The Tricks of Astrological Quacks. III. Of the Depth of the Sea. IV. Of Tobacco. V. Of Europes being too full of People. VI. The various Opinions concerning the Time of Keeping the Sabbath, London: Randal Taylor, 1688.

Lorraine Daston, "Attention and the Values of Nature in the Enlightenment," in Lorraine Daston and Fernando Vidal (eds.), *The Moral Authority of Nature*, Chicago: University of Chicago Press, 2004, 100–126.

Martin Davis and Virginia Davis, "Mistaken Ancestry: The Jacquard and the Computer," *Textile* 3, 1 (2005): 76–87.

Daniel C. Dennett, "Toward a Cognitive Theory of Consciousness," in *Brainstorms: Philosophical Essays on Mind and Psychology*, Cambridge, MA: MIT Press, 1978.

Daniel C. Dennett, "Will AI Achieve Consciousness? Wrong Question," *Wired* (February 19, 2019).

Carolyn Dicey Jennings, *The Attending Mind*, New York and Oxford: Oxford University Press, 2020.

Kenelm Digby, *Discours fait en une célèbre assemblée touchant la guérison des playes et la composition de la poudre de sympathie*, Paris: Augustin Courbé et Pierre Moet, 1658.

James Essinger, *Jacquard's Web: How a Hand-Loom Led to the Birth of the Information Age*, New York and Oxford: Oxford University Press, 2004.

Robert Fage, *A Description of the Whole World, with Some General Rules Touching the Use of the Globe*, London: J. Owsley, 1658.

Mohammed A. Farag, Huiming Zhang, and Choon-Min Ryu, "Dynamic Chemical Communication between Plants and Bacteria through Airborne Signals: Induced

Resistance by Bacterial Volatiles," *Journal of Chemical Ecology* 39 (2013): 1007–1018.

Orlando Figes, *The Europeans: Three Lives and the Making of a Cosmopolitan Culture*, New York: Metropolitan, 2020.

Michael Friedman, "Baroqueian Folds: Leibniz on Folded Fabrics and the Disruption of Geometry," in B. Sriraman (ed.), *Handbook of the Mathematics of the Arts and Sciences*, Cham: Springer, 2020, 1–28.

Jonardon Ganeri, *Attention, Not Self*, New York and Oxford: Oxford University Press, 2017.

William H. Gass, *A Temple of Texts: Essays*, New York: Alfred A. Knopf, 2006, 122–123.

Nicolas Grollier de Servière, *Recueil d'ouvrages curieux de mathématiques et de mécanique, ou Description du cabinet de M. Grollier de Servière*, Lyon: D. Forey, 1719.

Serge Gruzinski, *Quelle heure est-il là-bas? Amérique et Islam à l'orée des temps modernes*, Paris: Seuil, 2008.

Ian Hacking, "Do We See through a Microscope?" *Pacific Philosophical Quarterly* 62 (1981): 305–322.

Ian Hacking, *Historical Ontology*, Cambridge, MA: Harvard University Press, 2004.

Ellen Harlizius-Klück, *Weberei als episteme und die Genese der deduktivien Mathematik*, Berlin: Ebersbach, 2004.

Martin Heidegger, "The Question Concerning Technology," in *Basic Writings*, David Farrell Krell (ed.), New York: Harper & Row, 1977 [1954].

William James, *Principles of Psychology*, New York: Henry Holt & Company, 1890.

Hilton F. Japyassú and Kevin N. Laland, "Extended Spider Cognition," *Animal Cognition* 20 (2017): 375–395.

Immanuel Kant, *Critique of the Power of Judgment*, trans. Paul Guyer and Eric Matthews, Cambridge: Cambridge University Press, 2000.

Vera Keller, "Drebbel's Living Instruments, Hartmann's Microscosm, and Libavius's Thelesmos: Epistemic Machines Before Descartes," *History of Science* 48 (2010): 39–74.

Min-Shik Kim and Kyle R. Cave, "Perceptual Grouping via Spatial Selection in a Focused-Attention Task," *Vision Research* 41, 5 (March 2001): 611–624.

Friedrich A. Kittler, *Discourse Networks 1800/1900*, trans. Michael Metteer, with Chris Cullens, Palo Alto, CA: Stanford University Press, 1990 [1985].

S. Korsakof [Semyon Korsakov], *Aperçu d'un procédé nouveau d'investigation*, St. Petersburg, 1832.

Ray Kurzweil, *The Singularity Is Near*, New York: Viking, 2005.

Diogenes Laertius, *Lives of Eminent Philosophers*, vol. 2, Books 6–10, trans. R. D. Hicks, Loeb Classical Library, 185, Cambridge, MA: Harvard University Press, 1925.

Ganaele Langlois, "Distributed Intelligence: Silk Weaving and the Jacquard Mechanism," *Canadian Journal of Communication* 44, 4 (2019): 555–566.

Jaron Lanier, *You Are Not a Gadget: A Manifesto*, New York: Penguin Random House, 2011.

Jaron Lanier, *Ten Arguments for Deleting Your Social Media Accounts Right Now*, New York: Henry Holt and Company, 2018.

Frank La Rue, *Report of the Special Rapporteur on the Promotion and Protection of the Right to Freedom of Opinion and Expression*, Human Rights Council, Seventeenth Session Agenda, item 3, United Nations General Assembly, May 16, 2011.

Bruno Latour, "Why Has Critique Run Out of Steam? From Matters of Fact to Matters of Concern," *Critical Inquiry* 30 (2004): 225–248.

Antoine Le Grand, *Institutio philosophiae secundum Principia D. Renati Descartes*, Nuremberg: Johannes Zieger, 1679.

G. W. Leibniz, *Sbornik pisem i memorialov Leibnitsa otnosyashchikhsya k Rossii i Petru Velikomu*, ed. V. I. Ger'e, Saint Petersburg, 1873.

G. W. Leibniz, *Die philosophischen Schriften von G. W. Leibniz*, 7 vols., ed. C. I. Gerhardt, Berlin: Weidmann, 1875–90.

G. W. Leibniz, *Sämtliche Schriften und Briefe*, Berlin-Brandenburgische Akademie der Wissenschaften: Berlin/Potsdam, 1923—.

G. W. Leibniz, "Machina arithmetica in qua non additio tantum et subtractio sed et multiplicatio nullo, divisio vero paene nullo animi labore peragantur," trans. Mark Kormes, in David Eugene Smith, *A Source Book in Mathematics*, New York: Dover, 1959 [1685], 173–181.

Aldo Leopold, *A Sand County Almanac*, New York: Random House, 1966 [1949].

Rhodri Lewis, *Language, Mind and Nature: Artificial Languages in England from Bacon to Locke*, Cambridge: Cambridge University Press, 2007.

Tristan L'Hermite (ed.), *Recueil de pieces en prose, les plus agréables de ce temps. Composées par divers autheurs*, Part One, Paris: chez Charles de Sercy, 1659–1663.

Wenchao Li, ". . . 'früh bestelle, und etwas hinein bringe, so geld bringe!' Leibniz' Garten in Hannover," in Joachim Wolschke-Buhlman (ed.), *Gartenkultur im Spannungsfeld zwischen Arkadien und Soldatenfriedhöfen*, Munich: Thomas Martin Verlagsgesellschaft, 2015.

Franciscus Linus, *Tractatus de corporum inseparabilitate*, London, 1661.

Ada Lovelace, "Notes by the Translator" on L. F. Menabrea, "Sketch of the Analytical Engine invented by Charles Babbage," in Richard Taylor (ed.), *Scientific Memoirs, selected from the Transactions of Foreign Academies of Science and Learned Societies*, London: Richard and John Taylor, 1843.

Lucian, *A True History*, trans. Francis Hickes, London: A. H. Bullen, 1902 [1894].

Hermann Lotze, *Geschichte der Aesthetik in Deutschland*, Munich: G. J. Cotta'sche Buchhandlung, 1869.

Barry C. Lynn, "The Big Tech Extortion Racket: How Google, Amazon, and Facebook Control Our Lives," *Harper's Magazine*, September, 2020. https://harpers.org/archive/2020/09/the-big-tech-extortion-racket/

Stefano Mancuso, *The Revolutionary Genius of Plants: A New Understanding of Plant Intelligence and Behavior*, New York: Atria Books, 2017.

Robert McFarlane, "The Secrets of the Wood Wide Web," *The New Yorker*, August 7, 2016.

Marshall McLuhan, *Understanding Media: The Extensions of Man*, Cambridge, MA: MIT Press, 1994 [1964].

Luigi F. Menabrea, "Sketch of the Analytical Engine invented by Charles Babbage, Esq.," in Richard Taylor (ed.), *Scientific Memoirs, selected from the Transactions of Foreign Academies of Science and Learned Societies*, London: Richard and John Taylor, 1843.

Richard Menary, *The Extended Mind*, Cambridge, MA: MIT Press, 2010.

Stephen Menn and Justin E. H. Smith, *Anton Wilhelm Amo's Philosophical Dissertations on Mind and Body*, New York and Oxford: Oxford University Press, 2020.

Franz Mesmer, *Mémoire sur la découverte du magnétisme animal*, Geneva and Paris: P. Fr. Didot, 1779.

Carlos Montemayor and Harry Haroutioun Haladjian, *Consciousness, Attention, and Conscious Attention*, Cambridge, MA: MIT Press, 2015.

Henry More, *The Immortality of the Soul*, ed. A. Jacob, Dordrecht: Martinus Nijhoff, 1987 [1659].

Max Müller (trans. and ed.), *The Upanishads*, Part 2: *Brihadāranyaka Upanishad*, Oxford: Clarendon Press, 1887.

Vladimir Nabokov, *Pnin*, New York: Heinemann, 1957.

Antonio Pérez-Ramos, *Francis Bacon's Idea of Science and the Maker's Knowledge Tradition*, Oxford: Clarendon Press, 1988.

Michele D. Piercey-Normore and Christopher Deduke, "Fungal Farmer or Algal Escorts: Lichen Adaptation from the Algal Perspective," *Molecular Ecology* 20, 18 (September 2011): 3708–3710.

G. N. Povarov, "Machines for the Comparison of Philosophical Ideas," in G. Trogemann, A. Nitussov, and W. Ernst (eds.), *Computing in Russia: The History of Computing Devices and Information Technology Revealed*, Wiesbaden: Bertelsmann/Springer, 2001.

M. I. Radovskiĭ, "Iz istorii vychislitel'nykh ustroĭstv: 'intellektual'nye mashiny' S. Korsakova (po arkhivnym materialam AN SSSR)," in *Istoriko-matematicheskie issledovaniia*, vol. 14, Moscow: Fizmatizdat, 1961.

Agostino Ramelli, *Le diverse et artificiose machine del Capitano Agostino Ramelli*, Paris, 1588.

Colin Renfrew, "Neuroscience, Evolution, and the Sapient Paradox: The Factuality of Value and of the Sacred," *Philosophical Transactions B* 363, /499 (June 2008): 2041–2047.

Paul Ricoeur, *The Rule of Metaphor: Multi-Disciplinary Studies of the Creation of Meaning in Language*, trans. Robert Czerny with Kathleen McLaughlin and John Costello, London: Routledge and Kegan Paul, 1978.

Jessica Riskin, *The Restless Clock: A History of the Centuries-Long Argument Over What Makes Living Things Tick*, Chicago: University of Chicago Press, 2016.

Seph Rodney, "At This Year's São Paulo Biennial the Curator Prioritizes Feeling Over Discourse," *Hyperallergic*, September 20, 2018. https://hyperallergic.com/460749/at-this-years-sao-paulo-biennial-the-curator-prioritizes-feeling-over-discourse/

Simon Schaffer, "Babbage's Dancer and the Impresarios of Mechanism," in Francis Spufford and Jenny Uglow (eds.), *Cultural Babbage: Time, Technology and Invention*, London: Faber, 1996.

Susan Schneider, "It May Not Feel Like Anything to Be an Alien," *Nautilus* 080 (January 16, 2020).

James C. Scott, *Against the Grain: A Deep History of the Earliest States*, New Haven, CT: Yale University Press, 2017.

John Searle, "Minds, Brains, and Programs," *Behavioral and Brain Sciences* 3, 3 (1980): 417–457.

Michel Serres, "Internet, c'est vraiment du Leibniz sans Dieu," *Philosophie Magazine*, 21 August, 2012. https://www.philomag.com/articles/internet-cest-vraiment-du-leibniz-sans-dieu

Steven Shapin, *The Scientific Revolution*, Chicago: University of Chicago Press, 1996.

Brian Cantwell Smith, *The Promise of Artificial Intelligence: Reckoning and Judgment*, Cambridge, MA: MIT Press, 2019.

Justin E. H. Smith, "The Internet of Snails," *Cabinet Magazine* 58 (2016): 29–37.

Justin E. H. Smith, *Irrationality: A History of the Dark Side of Reason*, Princeton, NJ: Princeton University Press, 2019.

Roberto Garvia Soto, *Esperanto and Its Rivals: The Struggle for an International Language*, Philadelphia: University of Pennsylvania Press, 2015.

Manfred Stanley, *The Technological Conscience: Survival and Dignity in an Age of Expertise*, New York: Free Press, 1979.

William Sutton and John Sutton, "The Recording Sponge," *Fortean Times* 171 (2003): 56–57.

Atsushi Tero et al., "Rules for Biologically Inspired Adaptive Network Design," *Science* 22 (January 2010): 439–442.

Tom Vanderbilt, *You May Also Like: Taste in an Age of Endless Choice*, New York: Vintage, 2017.

Friedrich Theodor Vischer, *Aesthetik oder Wissenschaft des Schönen*, 6 vols. Reutlingen and Leipzig: Carl Mäcken's Verlag, 1846.

Robert Vischer, *Über das optische Formgefühl. Ein Beitrag zur Ästhetik*, Dissertation, University of Tübingen, 1872.

Michael Walzer, "Town Meetings and Workers' Control: A Story for Socialists," in *Radical Principles: Reflections of an Unreconstructed Democrat*, New York: Basic Books, 1980, 273–290.

Leslie White, "Energy and the Evolution of Culture," *American Anthropologist* New Series 45, 3, Part 1 (July–September 1943): 335–356.

Matthew D. Whiteside et al., "Mycorhizal Fungi Respond to Resource Inequality by Moving Phosphorus from Rich to Poor Patches Across Networks," *Current Biology* 29, 12 (June 17, 2019): 2043–2050.

Norbert Wiener, *Cybernetics: Or, Control and Communication in the Animal and the Machine*, Paris: Hermann & Cie., 1948; second edition, *Cybernetics, or Control and Communication in the Animal and the Machine*, Cambridge, MA: MIT Press, 1975.

James Williams, *Stand Out of Our Light: Freedom and Resistance in the Attention Economy*, Cambridge: Cambridge University Press, 2018.

Alan F. T. Winfield, "Experiments in Artificial Theory of Mind: From Safety to Story-Telling," *Frontiers in Robotics and AI*, June 26, 2018.

Tim Wu, *The Attention Merchants: The Epic Scramble to Get Inside Our Heads*, New York: Knopf, 2016.

Frances A. Yates, *The Art of Memory*, London: Routledge and Kegan Paul, 1966.

INDEX

A NOTE ON THE TYPE

This book has been composed in Arno, an Old-style serif typeface in the classic Venetian tradition, designed by Robert Slimbach at Adobe.